The Fundamentals of
Radiation
Thermometers

Peter Coates
Formerly of the National Physical Laboratory, UK

David Lowe
National Physical Laboratory, UK

CRC Press
Taylor & Francis Group
Boca Raton London New York

CRC Press is an imprint of the
Taylor & Francis Group, an **informa** business

CRC Press
Taylor & Francis Group
6000 Broken Sound Parkway NW, Suite 300
Boca Raton, FL 33487-2742

First issued in paperback 2019

ISBN-13: 978-1-4987-7821-3 (hbk)
ISBN-13: 978-0-367-88973-9 (pbk)

Library of Congress Cataloging-in-Publication Data

Names: Coates, Peter, -2013, author. | Lowe, David, 1962- author.
Title: The fundamentals of radiation thermometers / Peter Coates and David
Lowe.
Description: Boca Raton, FL : CRC Press, Taylor & Francis Group, [2016] |
Includes bibliographical references and index.
Identifiers: LCCN 2016022636| ISBN 9781498778213 (hardback ; alk. paper) |
ISBN 1498778216 (hardback ; alk. paper) | ISBN 9781498778220 (e-book) |
ISBN 1498778224 (e-book)
Subjects: LCSH: Radiation pyrometers. | Temperature measurements. |
Radiation--Measurement.
Classification: LCC QC277 .C63 2016 | DDC 681/.2--dc23
LC record available at https://lccn.loc.gov/2016022636

Visit the Taylor & Francis Web site at
http://www.taylorandfrancis.com

and the CRC Press Web site at
http://www.crcpress.com

Contents

CHAPTER 5 ■ Detectors   103

CHAPTER 6 ■ Series expansion analytical technique   135

List of Figures

List of Tables

Foreword

We are former colleagues and friends of the late Peter Coates at the National Physical Laboratory, and we were aware that he had more than twenty years ago brought to near-completion a book on the basic principles and experimental practice of Radiation Thermometry. Unfortunately at the time he was unable to complete and publish the work. Peter was an acknowledged and world-renowned expert in this field, and on reading the text it was clear to us that the book ought to be published. We are fortunate that Dr. David Lowe of the National Physical Laboratory has undertaken the considerable task of thoroughly revising the manuscript by removing techniques which had become outdated, adding new material, finishing text left by Peter in note form and preparing diagrams. We believe that the book will be illuminating to all practitioners who need to understand the subject, and warmly commend it to them. Finally, we thank Peter's widow, Sally, for giving this enterprise her full approval and encouragement.

Dr. B.P. Kibble
Dr. R.L. Rusby
Dr. D.J.E. Knight

Preface

This book emerged from notes and drafts written by Peter Coates around 1990, a draft for a comprehensive review of radiation thermometry as it stood at that time. These came to light after Peter's death in 2014, when his friend and former colleague at the National Physical Laboratory, Bryan Kibble, was dealing with his estate. Bryan contacted Richard Rusby, still at NPL, who asked me to look over the notes. I realised how useful they would have been to me during my 15 years at NPL. A number of times I have been involved in the specification, design and construction of radiation thermometers. As the UK national measurement institute, NPL does not manufacture mass production thermometers, but does get involved with making specific, niche instruments. There was a realisation that had I read what became this book earlier, it would have made my life easier and the instruments I was involved in making would have been better, or at least development would have gone more smoothly. It was clear to me that although unfinished, the contents were a valuable resource and I was eager to see them made available to my colleagues.

Peter worked at NPL in a number of fields. For many years he worked in radiation thermometry where he succeeded Terry Quinn as the section head of the thermometry group. He worked, amongst other things, on the NPL primary photoelectric pyrometer and made improvements to the stability of lamps used as reference standards. He is, though, perhaps best known for his work on multi-wavelength radiation thermometry, showing how to avoid the potential pitfalls of using this approach without adequate knowledge of emissivity. This is a theme that will be expanded in this book.

Peter obviously had a wide and deep knowledge of the physics and maths involved in non-contact thermometry. He had started to write a comprehensive overview of the entire field of radiation thermometry but with changes in his career and then retirement he had never completed the work. Some material was no longer relevant, for example, an entire chapter on the use of tungsten ribbon lamps. However a lot was still relevant, was written in a style that made it readable, and ideas were presented in a new way. When I had looked at what chapters I was reasonably confident I could complete, there seemed a good core of work with a coherent theme of the underlying processes and design considerations for making a measurement of temperature using Planck's radiation law. So when Bryan Kibble suggested a book, I agreed.

For someone working at a national measurement institute (NMI) it is tempting to write from the perspective of calibration and standards. NMIs tend to have fairly well-defined needs for their radiation thermometers, but of course most people do not work at NMIs. I tried to keep the work as general and non-specific as possible, the intent being to produce a resource for anyone intending to measure the temperature of an object using the radiated energy from that object. The first consideration is what we actually mean by the word temperature. There are some subtleties about the way temperature is defined and the way temperature scales are actually set up that are important to understand, and these are dealt with in the first chapter. There is also a brief look at the development of temperature scales that helps to understand why things are the way they are. Following this is a discussion about Planck's radiation law and the historical context in which it was derived. Since radiation thermometry actually measures the combination of temperature and emissivity it is important to understand the radiative properties of real materials. This is because emissivity is a complicated quantity that depends not only on the bulk property of the material being measured but also its surface condition and how that changes with thermal treatment. It is therefore important to know not only how reliable an emissivity value might be, but how reliable extrapolation to different wavelengths and different temperatures might be. Without this understanding it becomes impossible to know when a valid measurement with realistic uncertainties has been made, or if an inappropriate emissivity value has been used with consequent unknown errors. Chapters 4 and 5 deal with the 'hardware', the actual components that are required and how they interact, allowing a user to specify an appropriate design for a particular measurement problem. Peter developed a method for analytically treating the solution of Planck's radiation law. These days, of course, numerical techniques can readily be applied, but the 'Series expansion technique' covered in Chapter 6 allows for rapid analysis of small errors on the determined temperature. Chapter 7 covers an area where Peter was an acknowledged world leader: ratio and multi-wavelength radiation thermometry. Using two or more wavelengths is often suggested as a way to overcome problems with unknown emissivity. The potential pitfalls of this approach are considered, and situations where it is appropriate to use more than one wavelength are discussed. The problem of dealing with unknown emissivity often requires choosing a suitable approach to deal with a specific measurement situation. The final chapter looks at various strategies that can be used to minimise the uncertainties from unreliable emissivity data.

I would like to thank all those who contributed time and help, and their input has helped make this something of which Peter hopefully would have approved. In particular, and in no special order, Gavin Sutton, Martin Dury and Richard Rusby. Peter Quested was always available to give advice and it was appreciated. Finally, I would not have done this work without the prompting of Bryan Kibble.

Authors

Peter Coates joined the National Physical Laboratory (NPL) in 1966, having earned a BSc at Cambridge and a PhD at Imperial College working with shock tubes. Upon joining the NPL, he initially constructed ultra-fast timing circuits for measuring the time of emission of radiation from atoms, a skill he took with him when he joined the Temperature Section to measure the emission of photons from poorly radiating surfaces by counting individual photons. He transferred to the Temperature Section in 1972 to apply his expertise in photon counting to the NPL primary photoelectric pyrometer, built by T J Quinn and M Ford. He was able to make significant improvements in precision use of photomultipliers in the days before silicon photodiodes became established. He and his colleagues, Terence Chandler and John Andrews, also improved the performance and use of pyrometric lamps, including feedback stabilisation of the radiance, and they made the first and most accurate determination of the freezing temperature of palladium in many years. Dr. Coates succeeded Dr. Quinn as Section Head in 1975, while continuing his work in radiation thermometry. He was an excellent theorist, and produced two or three seminal papers in radiation thermometry, notably exposing the weaknesses of multi-wavelength methods, which had been much trumpeted as overcoming the difficulty of unknown emissivity, showing that it is fundamentally based on an unjustified extrapolation to zero wavelength and that passive techniques alone could not solve the problem. In the early 1980s, dissatisfied with the frustrations of management, he made a second career change and moved to the NPL Time and Frequency Group, where he remained until his retirement.

David Lowe joined NPL in 2000, having earned a BSc in physics at University of Wales, Cardiff, and a PhD in engineering at Warwick, working on optical methods to characterise semiconductor surfaces. His first job at NPL was building a primary radiation thermometer for realisation of ITS-90 above the silver point. He has worked on the development of high temperature fixed points, and has built a number of radiation thermometers from 5 μm to 500 nm operating wavelength.

CHAPTER 1

The quantity 'temperature'

1.1 INTRODUCTION

Radiation thermometry[1] may be defined as that field of science which deals with the measurement of the quantity 'temperature' using the characteristics, such as the intensity or the spectral distribution, of the thermal radiation emitted by the object undergoing measurement. The first step, then, is to be clear what quantity is being measured and how it should be expressed. In particular, we should be clear of the distinction between an empirical scale and thermodynamic temperature. It is the aim of this chapter to provide a brief introduction to the quantity 'temperature', and set the context of its measurement, before proceeding to discuss the fundamentals of making a measurement of radiant intensity (or intensities) and from there assigning a temperature value.

1.2 THE CONCEPT OF A SCALE

Objects and events in the world around us show a variety of characteristics or properties. Some of these possess magnitude, that is, some of the objects have the property to a greater or lesser extent than others. These properties, for example size and weight, are called quantities; shape and colour, on the other hand, are generally considered to be qualities. One of the basic assumptions of science, and one reason why it has proved so successful in describing the physical world, is that it is possible to measure quantities.

Measurement is the assignment of a number according to established objective rules or procedures, in such a way that the number represents the magnitude of the quantity concerned.

Quantities generally have a continuous range of values and cannot be determined exactly. There is always some uncertainty associated with the measurement of a

---

[1]A note on terminology: Historically the term 'pyrometer' has been widely used. At the present time the term 'radiation thermometer' is more generally favoured. In this book we use the term 'radiation thermometer' as a general term for an instrument of any description that uses the emitted radiance of an object as a means of determining that object's temperature. It is not restricted to any particular spectral band, nor the number of wavelengths used, nor the design of the instrument itself.

physical quantity, and this greatly complicates both the theory of measurement and its application in the real world. In addition, the units in which physical quantities are measured characterise the quantity itself and are independent of the application. This is not to say that different units may not have been adopted for a given quantity, but the relationship between them is usually well known or even laid down by definition. The length of a molecule may be given in Angstrom units, that of a cricket pitch in yards, and the distance to the moon in kilometres, but all may be expressed if need be in metres.

Since in the physical sciences the fundamental concepts are well defined and the measurements of the corresponding quantities are limited only by the uncertainties associated with the measuring procedures, it is possible to state the conditions necessary for the existence of a physical quantity quite positively.

In the discussion which follows we restrict ourselves to scalar, as opposed to vector, quantities. The conditions required for a property of a system to be recognised as a quantity may be summarised as follows. First, the property must possess physical or theoretical significance. It is possible to invent quantities for which the following rules will apply, but which have no real meaning. For example, it is possible to assign quite accurate values to a variable which is formed by taking the product of a person's height and age (their 'hage', measured in *metre.years*), but such a quantity is at present of no value whatsoever. The second requirement is that it shall be possible to find procedures which allow all of the objects or events that demonstrate the property to be ordered, that is, arranged in order of magnitude.

The ordering of objects in general is quasi-serial, that is, objects are divided into groups, or sub-sets, in which each member possesses the same magnitude of the quantity, and the groups are then arranged in ascending order. The next step is to establish a procedure which assigns to each sub-set a number. For physical quantities, the density of sub-sets is extremely high, and we may treat the quantity as essentially taking a range of continuous values. The range is not necessarily the whole range of real numbers, as many quantities are limited to positive values.

It is important to realise that the procedures used do not imply any relationships between the numbers assigned in different parts of the permitted range, that is, the suggestion that such scales must be 'linear' in some way is meaningless. In addition, the magnitude of the unit interval, that is, the difference between two states whose assigned numbers differ by unity, cannot necessarily be related in different parts of the scale.

For a quantity without a solid theoretical background, scales based upon different operational definitions will give arbitrary values which are not necessarily related in any predictable way. One of the first indications that a quantity has real physical significance and is well understood is that it becomes apparent that there are simple numerical relationships between scales based on a variety of physical effects. Often, values on one scale may be converted to another with a linear transformation. Initially, the relationships may be obscured by experimental uncertainties and the presence of unsuspected systematic errors. But eventually, when the structure of the theory is sufficiently well established, it becomes possible to jettison the empirical system of arbitrary scales, and to replace it with a system of measurement based upon a standard amount of the quantity, known as the unit quantity. Different magnitudes of the quantity may then be built up from multiples and sub-multiples of the unit quantity, using any of the established relationships within the theoretical structure. The major advantage of adopting this approach is that of generality, as the concepts involved may be applied equally to new physical laws and experimental techniques, as the subject area develops and the theoretical superstructure is extended.

The adoption of a system of unit quantities removes the dependence on operational definitions of scales which must specify particular procedures and materials to be used in the measurement process, and allows the application of any of the physical laws which involve the quantities concerned. However, the point of contact with the real world then occurs in the need to be specific about the definition of the unit quantity itself. In classical physics the units or primary standards were realised in the form of material artifacts, such as the platinum-iridium metre bar and the kilogram. It was necessary to treat them with great care, and to bring them out for comparison with secondary or national standards on relatively rare occasions. Although the system was not so fragile that an accident to the primary standard would have led to a significant change in the unit quantity, it was extremely difficult to ensure that slow drifts in their absolute magnitude did not occur.

More recently some of the definitions of units have been changed, or are in the process of being changed, so that they refer to atomic rather than macroscopic properties. Of course, this has to be carried out without introducing a significant change in the magnitude of the unit. The constants of the atomic processes concerned should be at least as well known as the material standards they are proposed to replace. This approach has two very significant advantages. First, it is a fundamental tenet of quantum theory that atoms of a particular isotope and in a given state are not simply similar in their characteristics but identical. The unit quantity may be realised independently in laboratories around the world and should in principle be the same in each. It is therefore much more accessible than a single artifact, and much less prone to accidental change. In practice, of course, the distinction is blurred by the effects of systematic uncertainties which differ from one laboratory to the next. This leads to the establishment of systems for inter-comparing the national standards in order to assess the overall reproducibility of the unit. The second advantage is that there is no reason to suppose that atomic properties vary with time. Although some theories suggest that fundamental constants may change on a cosmological time scale, the effects are limited, very small in magnitude and have not been confirmed in practice.

1.3 THERMOMETRY – THE MEASUREMENT OF TEMPERATURE

The concept of the temperature of an object may be directly experienced as part of our everyday existence, but measurement of the temperature of an object by touch is a very subjective process and also depends on the properties of the object itself, in particular its thermal conductivity. What is sensed, of course, is the heat that flows from the object to, or from, the sensors in our fingers or hand. But temperature and heat are different quantities, and they are commonly distinguished by saying that "temperature is the degree of hotness," though exactly what this means is not clear. The most general definition of temperature comes from the first law of thermodynamics[2], which considers heat flow between objects and establishes the idea of 'thermal equilibrium'. It can be stated as "Temperature is the potential for heat flow," or, more generally, "the potential for heat transfer" by conduction, convection or radiation. Thus a temperature difference causes heat to flow, much as a pressure difference causes water to flow in a pipe or an electric potential difference causes a current to flow in a wire.

Returning to our everyday sensation of temperature, the range of temperature which can be sensed by direct contact is very limited. At freezing temperatures, and those above about 60 °C, physical contact becomes increasingly painful. As an alternative, the temperature of hot objects may be sensed by non-contact means: the thermal (heat) radiation may be detected by the warming sensation produced. When the source is at a very high temperature, enough radiation may be emitted at visible wavelengths to be detected by the eye. The apparent colour moves from a dull red to an intense white or even blue as the temperature increases, and this gives rise to such qualitative estimates of temperature as 'dull red heat', 'red hot' and 'white hot'.

The limited range and accuracy of human perception was of course the main stimulus for the development of thermometers, that is, instruments to measure temperature quantitatively and not just qualitatively. One of the respects in which temperature is different from other common quantities is that it is 'intensive', that is, its magnitude is independent of the amount of material in the object or system concerned. Mass and volume, on the other hand, are extensive quantities; if we take a block of copper, for example, and cut it into two equal parts, the mass and volume of each is half the original value, but the temperature remains unchanged. Partly as a result, temperature is always measured indirectly, in terms of a property of the thermometer which varies with temperature and is easily and accurately measured. This form of measurement is termed 'associative'. A numerical relationship must be assumed or derived in order to convert the measurement of the associated quantity into a value for the temperature. The first effective thermometers were developed in Italy around 1641 as a result of the skills of Florentine glass blowers. The thermometers consisted of sealed alcohol-in-glass instruments; some, constructed with long coiled stems, were sensitive enough to respond to a human breath. The stems were marked with a scale which had divisions equal to a constant fraction of the volume of the bulb. In 1664, Hooke, then Curator of the Royal Society, obtained a Florentine thermometer. He modified the design, introducing a red dye into the

---

[2]The second law can also be used, however, this can lead to the concept of 'negative' temperature with the odd property that objects with negative temperature are hotter than any object with positive temperature.

alcohol to make the liquid column more visible. While the Florentines had made their thermometers with characteristics as uniform as possible in order to achieve reproducibility, Hooke defined a scale of more general applicability. He took the freezing point of water as zero, and made each degree equal to a change of one five-hundredth of the volume of the alcohol.

The next major advance was the discovery by Fahrenheit around 1724 of the methods needed to make reliable mercury-in-glass thermometers. He also adopted the idea of the Danish astronomer Romer of generating a scale based upon two fixed points, the freezing point of water and the temperature of blood in the human body, and of evenly dividing the interval between them into a convenient number of degrees. To ensure that all common temperatures were positive, the zero was set at the lowest air temperature known. It was found that this was below the ice point by about one-half of the interval between the fixed points. Thus Fahrenheit took numerical values of 32 °F for the ice point and 96 °F for that of blood. Celsius, a Swedish astronomer, introduced in 1742 a form of the centigrade scale, with one hundred degrees between the freezing and boiling points of water. Initially the usual assignment of values was inverted, that is, the ice point was 100 °C and the steam point 0 °C, but his colleague Strömer soon reversed the order to produce what has become the most widely used temperature scale.

These early scales differed from one to another not only in that the numerical values for a particular temperature varied, but in a more subtle fashion. The scales were divided linearly, that is, the distance between the fixed points on the stem of a thermometer was divided uniformly into the required number of degrees. Reaumur realised in 1739 that the scales of mercury and spirit thermometers differed because the volume expansion coefficient of the two liquids is not constant at different temperatures. Clearly, each type of thermometer realises a different scale.

1.4   THERMODYNAMIC TEMPERATURE

Soon after 1700 an alternative approach was initiated by the work of Amontons, who related temperature linearly to the pressure of a fixed volume of air. He found that the extreme summer and winter temperatures in Paris corresponded to pressures in the ratio 6:5. Conjecturing that there was a minimum temperature achieved when the pressure of air fell to zero, he suggested that if this were taken as zero, only one fixed point would be needed to define the scale completely. This prescient proposal was eventually adopted over 250 years later. Later experiments on the properties of gases by Charles, Dalton, Gay-Lussac and Regnault around the end of the eighteenth century led to the discovery that the coefficient giving the change of pressure with temperature was nearly the same for all gases. The advantages of defining a temperature scale which was independent of the working fluid were recognised by Dulong and Petit in 1817. It was later shown that a centigrade scale established on a constant volume gas thermometer by calibration at the freezing and boiling points of water was very close to that established on a mercury-in-glass thermometer with the same fixed points.

The theories of ideal gas behaviour became incorporated during the nineteenth century into the much more general and powerful structure of thermodynamics. The concept of temperature formed a fundamental part of this, and its relationship with other thermal quantities became more clearly understood. It is possible to construct from many physical effects a set of state transformations which are, for illustrative and historical reasons, known as heat engines. Their function is to transform heat energy and work through a cycle of operations that returns them to their initial condition. Using very simple arguments, it is possible to show that the efficiency η of a reversible heat engine depends only upon the temperatures of the sink and the source between which it operates, that is,

$$\begin{aligned} 1 - \eta &= F(T_A, T_B) \\ &= \frac{f(T_A)}{f(T_B)} \end{aligned} \tag{1.1}$$

The efficiency is a maximum for given values of T_A and T_B and does not depend upon the underlying mechanism or the materials involved.

In 1848 Kelvin proposed that these ideas could form the basis of a theoretical definition of temperature which was much more general than the existing empirical scales. By adopting a linear function

$$f(T) = cT \tag{1.2}$$

the thermodynamic temperatures[3] obtained could be shown to be identical to those determined by gas thermometry. The calculation involved a heat engine based upon an ideal gas undergoing a Carnot cycle, a sequence of adiabatic and isothermal expansions and contractions. As might be expected, the efficiency η can only vary between zero and unity. From (1.2) above, temperatures are thus limited to positive values only. Indeed the lower limit, the absolute zero of temperature, is shown to be inaccessible as it corresponds to a perfectly efficient engine with $\eta = 1$. The heat flow into the sink would then be zero, and it would be possible to construct an engine which completely transformed heat into work, an effect prohibited by the second law of thermodynamics.

To assign numbers to the thermodynamic temperatures produced, it is necessary to define the constant of proportionality c in (1.2). The simplest technique is to assign a value to a convenient and reproducible fixed point. When this approach was adopted in 1956 the triple point of water was chosen, with a thermodynamic

---

[3]At this point we should mention another problem in terminology. Some authors employ the qualifiers 'primary' and 'secondary' when applied to thermometry as being equivalent to the terms 'thermodynamic' and 'empirical', as defined above. This usage clashes with that used in the calibration of thermometers. A primary standard thermometer is one which has been calibrated by direct reference to an accepted thermodynamic method or an International Temperature Scale, neither of which require temperatures determined from any other source. Secondary standards may then be obtained either by direct comparison with primary standard thermometers, or by using fixed points or other techniques whose values have been established using the primary thermometers.

temperature of 273.16 K. At this time, the formal definition was phrased in terms of a thermodynamic temperature scale and contained no mention of a temperature unit. However, in 1968 the wording was changed to indicate that the choice of a value for the triple point of water effectively established the magnitude of the unit of thermodynamic temperature, the kelvin.

Although the concept of a temperature unit is more abstract than those of mass and length, for example, it is not in any way less meaningful. It has been noted that any physical law which depends upon temperature may be employed to realise thermodynamic temperatures, although some will be more convenient or accurate than others. Each technique would be expected, within its limitations, to give the same value for a given object or system. In addition, each law provides a demonstration of the fact that the unit of thermodynamic temperature does not in principle depend upon temperature – a proof which is impossible for temperatures measured upon empirical scales.

1.5   INTERNATIONAL TEMPERATURE SCALES

Having stressed at some length the significance of thermodynamic temperature, it may come as a surprise to learn that the vast majority of temperature measurements, including those in many scientific experiments, are made upon an empirical scale of temperature. The reasons are, upon reflection, fairly obvious. First, the number of fundamental physical laws upon which measurements of thermodynamic temperature may be based and which are also practical and capable of good accuracy is rather small, particularly at temperatures above 0 °C. Secondly, thermodynamic methods turn out to be rather difficult and cumbersome to apply, especially when the highest accuracy is required. Temperature scales, on the other hand, may be based on any convenient property which varies reproducibly with temperature, especially if it is easily measurable.

The remainder of this chapter will be devoted to a discussion of the philosophy and structure of the International Temperature Scales. Modern radiation thermometry covers temperatures down to and below the ice point, and in much of this range practical temperatures are defined in terms of the performance of a platinum resistance thermometer. The relationship between them requires an explanation. Above the freezing point of silver, 961.78 °C, the current ITS-90 employs a monochromatic radiation thermometer to establish the scale, and the conceptual difficulty largely disappears.

1.5.1   The origin of internationally agreed scales

The first stage in the realization that it was undesirable to promote the use of thermodynamic temperatures as the working standard for thermometry came with the unification, primarily for reasons of trade, of the system of measurement units via the Convention du Metre. The Bureau International des Poids et Mesures (BIPM) in Paris was to look after the new primary standards of the metre and the kilogram, and promote agreements on other units. The main requirement for thermometry at BIPM was to measure the temperatures of the platinum-iridium metre rods.

In the period around 1886, Chappuis at the BIPM compared thermodynamic temperatures obtained by constant volume and constant pressure gas thermometry, using hydrogen, nitrogen and carbon dioxide as the working fluid, with an empir-ical scale based on mercury in glass thermometers [1]. This scale was obtained by marking the thermometers at the freezing point and the boiling point of water, and linearly dividing the fundamental interval into one hundred degrees. The results of Chappuis' work, shown in Figure 1.1, were repeatable to better than one-hundredth of a degree. In 1887, it was decided as a result to adopt as a standard a limited scale based upon a constant volume gas thermometer using hydrogen. This was known as the normal hydrogen scale. As it was based upon the freezing and boiling points of water, and no correction was made for deviations from ideal gas behaviour, it differed slightly from thermodynamic temperature and was referred to as a practical scale. To ensure reliable results, it was necessary to specify the filling pressure of the gas thermometer at $0\,^{\circ}C$. The specification of fixed points, the nature of the gas and the filling pressure all form part of the operational definition, and we may conclude that the normal hydrogen scale is correctly termed a scale, even though the temperature values derived from it were very close to thermodynamic, within a few millikelvins.

The techniques of gas thermometry were extended in the next twenty years up to very high temperatures, but the development of the platinum resistance thermometer by Callendar provided a much more convenient and precise method of measuring temperatures to above $600\,^{\circ}C$. Although Callendar proposed in 1899 that a practical scale should be based on the platinum resistance thermometer as a British standard, it was not until 1927, partly as a result of World War I, that it was eventually adopted and formed the basis of an internationally accepted temperature scale.

By that time, the original concept had been significantly revised and extended as it had been found that resistance thermometers became unstable at very high temperatures, mainly as a result of the contamination of the platinum. The first International Temperature Scale, known as ITS-27 from the year of its acceptance by the CGPM[4], comprised three contiguous ranges[5] and thus had the form that

---

[4]The Conference Generale des Poids et Mesures, an international group of scientists meeting every four years at BIPM and empowered to authorise changes in the system of units, usually on the advice of the various Consultative Committees.

[5]Temperatures measured in kelvins are required for measurements made at very low temperatures and are known as 'Kelvin temperatures'. For applications at normal ambient temperatures and above, it is often more convenient to express them as 'Celsius temperatures' with a symbol $^{\circ}C$, where the zero has been moved to the freezing point of water, that is,

$$\frac{t}{^{\circ}C} = \frac{T}{K} - 273.15$$

As shown, Kelvin and Celsius temperatures are conventionally indicated by the use of upper- and lower-case letters, respectively. Values on different International scales are shown by the addition of

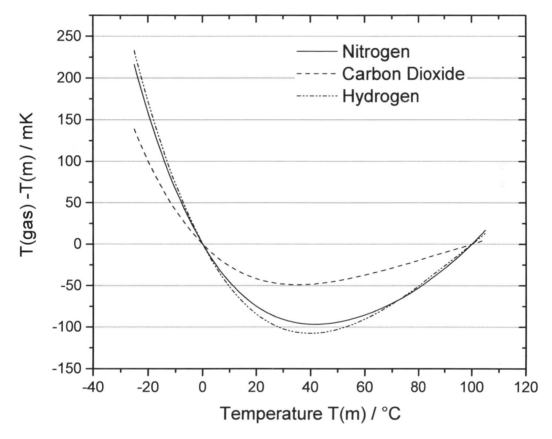

Figure 1.1: The differences between gas thermometry based on different gases and the scale generated on accurate mercury-in-glass thermometers.

continues to the present day of different specified instruments and interpolation functions for different temperature ranges. The difficulty is, of course, ensuring that what are effectively independent scales behave in an acceptable way where they meet. Not only should the two scales give the same temperature value at the intersection, but there should be no sudden change in the magnitude of the unit: the slope of the interpolation functions should be so far as possible equal.

The ITS-27 specification was:

- From $-190\,°C$ to $660\,°C$ it was defined using a platinum resistance thermometer. Above $0\,°C$, this was to be calibrated at the freezing point of water ($0\,°C$), the boiling point of water ($100\,°C$) and the boiling point of sulphur ($444.60\,°C$). The variation of the resistance R_t was given by the Callendar quadratic equation

$$\frac{R_t}{R_0} = 1 + At + Bt^2 \tag{1.3}$$

  where t was the temperature in degrees Centigrade (sic) and R_0 was the resistance of the thermometer at $0\,°C$. The characteristics of the thermometer, mainly determined by the purity of the platinum, were limited by specifying that the resistance ratio R_t/R_0 should not be less than 1.390 at the steam point, and not less than 2.645 at the boiling point of sulphur. Below $0\,°C$, the Callendar-van Dusen equation

$$\frac{R_t}{R_0} = 1 + At + Bt^2 + Ct^2(t - 100) \tag{1.4}$$

  was to be used, the extra (cubic) term being determined by calibration at the boiling point of oxygen ($-182.97\,°C$).

- Between $660\,°C$ and $1063\,°C$, a thermocouple, the positive arm of which was a 10% rhodium-platinum alloy, the negative pure platinum, was employed. This was to be calibrated at the freezing points of antimony ($630.5\,°C$), silver ($960.5\,°C$) and gold ($1063\,°C$), with a quadratic interpolation equation

$$E = a + bt + ct^2 \tag{1.5}$$

  to relate the measured thermoelectric voltage E to temperature.

- Above the gold point, temperatures were to be obtained by extrapolation using an optical pyrometer. This was assumed to obey the Wien radiation law, and the second radiation constant specified to be $1.432\,\mu m \cdot K$. The operating wavelength λ was limited by the requirement that the product $\lambda(t + 273)$ should be less than 0.3 cm.degrees.

It can be seen that there is no reason to suppose that the platinum resistance scale and thermocouple scale will match at the boundary at 660 °C. We will consider this following a short digression on the subject of 'wire' scales.

---

the last two digits of the year of introduction as a subscript. For example, t_{27} implies a temperature in degrees Celsius referred to as ITS-27.

1.5.2 Wire scales

One of the functions of the metrologists in the national laboratories is to provide primary standards for each quantity to ensure that measurements made around the world are compatible. In the case of temperature, we have seen that these might be referred to fundamental physical laws which are universal but require a great deal of time and skill in their realization, or upon empirical scales which are to some extent arbitrary in their definition but employ convenient and reproducible thermometers. There is in fact a third way in which a reference standard could be established which combines elements of both methods. For each temperature range, a group of stable thermometers would be selected, preferably from a single batch, so that their characteristics would be very similar. They would then be calibrated on one occasion with a thermodynamic method such as gas thermometry. The scale would thereafter be defined in terms of the characteristics of these particular thermometers. Such a scale is generally known as a 'wire scale', as the original examples employed resistance thermometers.

The system has the advantages that it is inexpensive to maintain and that the calibration of working standards, by comparison methods, is both simple and accurate. On the other hand, the set of thermometers must be held intact at a central laboratory, as their inter-comparison is a necessary part of maintaining the reliability of the standard. Other countries are required to send their standards to the centre for calibration; this may be inconvenient and even politically undesirable. There is always a danger that changes in calibration might occur in transit. While it is possible that an accident at the central laboratory might cause the loss of the scale, the chance of this is small. The main reason that this simple approach has not been accepted internationally is that it would be very difficult to detect a slow drift in the characteristics of the set of thermometers, so that the scale changes imperceptibly. However, groups of thermometers have been used as a 'flywheel' between the expensive realisations of a more fundamental scale, particularly at cryogenic temperatures where there are few reasons for calibration drifts.

The international temperature scales have the advantage that they can be set up from scratch in any laboratory with the desire, and the resources necessary, to do so. In an abstract sense, this means that they are relatively insensitive to accidental damage and unlikely to drift with time. In practice, it is very convenient for a technically advanced country, where thermometry at the highest levels of accuracy may be required, to have direct access to the primary standards. This is not to say that each realization of the scale should be completely independent. Collaboration and inter-comparison with other national standards laboratories not only reduce the technical problems and the possibility of blunders, but enable the reproducibility of the scale itself to be estimated.

1.5.3 The International Temperature Scale of 1990, ITS-90

The ITS-27 was succeeded by updated scales at various times. The issue of the platinum resistance thermometer and thermocouple scales not meeting at their boundary was addressed in 1948 by having different regions meet at a defined fixed point, giving the ITS-48. This was further refined in 1960 to take account of changes in the International System of Units (SI) with the kelvin defined in terms of the triple point of water to give the International *Practical* Temperature Scale of 1948, IPTS-48[6]. This in turn was revised to extend the range to lower temperatures, and to change the available reference fixed points and their temperatures with the IPTS-68. Finally, the current scale, ITS-90, lost the word 'Practical' in its name, lost the thermocouple as a specified instrument, further extended the range to lower temperature, and provided reference temperatures that made it a much better approximation to thermodynamic temperature.

Generally, multiple issues are addressed at each revision. A scale revision occurs when the accumulated issues make the current version less acceptable than an update despite the upheaval that the update inevitably brings. These issues include not just improved thermodynamic T measurements, but also changes in unit definition, the availability of easier-to-use fixed points and improved interpolation schemes. These are all currently active issues so after considering the ITS-90, which is the almost universally accepted standard for thermometry, we will briefly consider what might follow it.

ITS-90 [2][3] supersedes the IPTS-68 and a provisional low temperature scale, the 1976 Provisional 0.5 K to 30 K Temperature Scale (EPT-76), and it covers temperatures from 0.65 K upwards. Its structure above 0 °C will be described here as that is the temperature range covering most radiation thermometry. The defining fixed points of ITS-90 and the values assigned to them are given in Table 1.1. It will be seen that neither the ice nor the steam point is included, making this the first temperature scale completely divorced from the 'fundamental interval'. New fixed points, gallium (Ga), indium (In) and aluminium (Al), were introduced for the calibration of platinum resistance thermometers.

A major change from previous scales was the elimination of the thermocouple as an interpolation instrument because of its inherently poor reproducibility. The platinum resistance thermometer and radiation thermometry ranges were extended to fill the gap, but the selection of the junction temperature gave rise to some problems. Ideally, the resistance thermometer should have covered the complete thermocouple range, as the gold point is a very reproducible (if expensive!) fixed point. Unfortunately its stability deteriorates quite rapidly above about 950 °C. The radiation thermometer could be employed down to the aluminium point, but its lack of long-term stability and the associated difficulties of dissemination of the scale made this undesirable. It was therefore agreed that the junction between the two should lie at the silver point, although this is not the best fixed point for thermometric use as it is affected by the presence of oxygen. The new scale allows the use of either the silver, gold or copper points as the defining fixed point for radiation thermometry. It is claimed that the uncertainty in the recommended

---

[6]The temperatures remained the same on both scales, hence the year associated with the scale stayed as 1948.

values for these points is such that the temperatures derived will not differ by an amount important for any practical application.

Table 1.1: Defining fixed points of the ITS-90 above 0 °C

Substance	State	Kelvin temperature	Celsius temperature
H_2O	Triple point	273.16	0.01
Ga	Melting point	302.9146	29.7646
In	Freezing point	429.7485	156.5985
Sn	Freezing point	505.078	231.928
Zn	Freezing point	692.677	419.527
Al	Freezing point	933.473	660.323
Ag	Freezing point	1234.93	961.78
Au	Freezing point	1337.33	1064.18
Cu	Freezing point	1357.77	1084.62

Note: Melting and freezing points are defined at a pressure of 101 325 Pa.

Because the stability of conventional $25\,\Omega$ long-stem platinum resistance thermometers is poor at the silver point, a special low resistance version being required to reach this temperature, it was decided to allow the calibration to be performed over parts of the range to suit the thermometer design and intended usage. The measured quantity $W(T_{90})$ of the thermometer is now given as the ratio of the resistance at temperature T_{90} to that at the triple point of water, that is,

$$W(T_{90}) = \frac{R(T_{90})}{R(273.16)} \tag{1.6}$$

Thermometers suitable for use as interpolation instruments must have resistance ratios greater than specified limits at various temperatures, depending upon their intended range of use. This is a practical way of ensuring that the wire used is of suitable standard in such aspects as purity that are otherwise hard to assess. The reference function is given as a ninth-order polynomial

$$W_R(T_{90}) = \sum_{i=0}^{i=9} c_i \left[\frac{T_{90} - 754.15}{481} \right]^i \tag{1.7}$$

where the coefficients c_i are given in a table. The inverse function is also provided. Real thermometers, of course, will differ slightly from this reference polynomial and calibration involves determining the coefficients of a deviation function. Some of the coefficients of the deviation function

$$\begin{aligned} W_D(T_{90} &= W(T_{90}) - W_R(T_{90}) \\ &= a[W(T_{90}) - 1] + b[W(T_{90}) - 1]^2 + c[W(T_{90}) - 1]^3 \\ &\quad + d[W(T_{90}) - W(660.323\,°\mathrm{C}]^2 \end{aligned} \tag{1.8}$$

are set to zero for calibration over part of the range. Table 1.2 shows the recommended ranges and fixed points required.

Table 1.2: Recommended sub-ranges and fixed points of the ITS-90

Range	H$_2$0	Ga	In	Sn	Zn	Al	Ag	Coefficients
0 °C to Ag	●			●	●	●	●	
0 °C to Al	●			●	●	●		d=0
0 °C to Zn	●			●	●			c=d=0
0 °C to Sn	●		●	●				c=d=0
0 °C to In	●		●					b=c=d=0
0 °C to Ga	●	●						b=c=d=0

ITS-90 is the first temperature scale to allow a choice of realisations within each range. Strictly speaking, this means that it forms a collection of scales rather than a single scale as in the case of the earlier International Temperature Scales. It may be maintained that they can be regarded as a single scale if for all practical purposes they need not be distinguished. It is stated in the text accompanying the definition that each will have equal status and that the additional non-uniqueness of the scale will be of no practical importance. Information on the practical realization and the accuracy of the scale are provided in electronic form by the BIPM as "A Guide to the Realization of the ITS-90" [4]. In addition, two documents, 'Supplementary Information for the ITS-90' [5] and 'Techniques for Approximating the ITS-90' [6], published by the BIPM also contain useful information.

1.6   THE DEFINITION OF TEMPERATURE UNITS

One of the more confusing statements in the definition of the International Temperature Scales is that the unit of practical temperature is the same as that of thermodynamic temperature. It is not at all clear what is intended: taken at face value, it suggests that two objects, one with a temperature of 100 K measured on an International Scale, the other with a thermodynamic temperature of 100 K, will be in thermal equilibrium, as in both cases the number and the unit are identical. However, we know this is not true, as much time and effort has been, and continues to be, spent by scientists at the various national laboratories in determining the value of these differences, which are published by the BIPM. It is necessary to explain how the situation arose.

When the first international temperature scale, ITS-27, was set up, both it and measurements of thermodynamic temperature, such as those of Chappuis at BIPM, were based upon the 'fundamental interval', that is, on the 100-degree separation of the ice and steam points. This largely arose from the historical development of thermometers and thermometry. In 1954, the proposal to base the unit of thermodynamic temperature upon the interval between absolute zero and a single fixed point was at last accepted. The idea, as we have seen, was originally put forward by Amontons in 1702 and extended by Kelvin in 1854. In the event, the triple point of water was chosen, partly because it was close to the ice point, still felt by many to be the 'origin' of the scale, but also because it is easy to realise and very reproducible. The temperature assigned to it, 273.16 K, effectively defined the magnitude of the unit of thermodynamic temperature, the degree Kelvin.

When the international temperature scale was being revised before 1968, there was a feeling that the scale unit still depended upon the 'fundamental interval', and hence was defined in a completely different way from the thermodynamic unit. In addition, as the value assigned to the triple point had been determined experimentally from the fundamental interval, it was possible that their magnitudes were different. The solution arrived at was to 'define the problem away' by declaring that the two units were the same, and to stress that the same value was assigned to the triple point of water in both systems.

The argument, of course, contained a number of fallacies. It ignored the difference between the unit scale interval on an empirically defined temperature scale and the unit quantity relevant for the theoretically based thermodynamic temperature. In the definition of an International Scale, the triple point and boiling point of water carry no more weight than any of the other defining fixed points, and the unit cannot be regarded as being determined in any way by the fundamental interval. The only measure of the magnitude of the unit scale interval is to compare it with the thermodynamic unit, and we find that there are differences, often small but sometimes not so small, which vary with the position on the scale. This must be regarded as inevitable; however carefully the definition of a practical scale is prepared, it will differ in some degree from thermodynamic temperature, and the differences will vary along the scale. We must in fact conclude that the units are not only different in magnitude but fundamentally different in type.

In many cases the deviations between international temperature scale and thermodynamic temperature are negligible in magnitude, and these points are not so important, but for measurements pushing the limits of what is technologically possible it is vital to stress the differences and the deviations, as the present situation merely causes confusion. There are many users who do not appreciate the difference between practical and thermodynamic temperature, and in some cases it has affected their results.

1.7   THE MISE EN PRATIQUE OF THE DEFINITION OF THE KELVIN

It has been recognised that improvements to thermometry could allow an improved ITS, in the sense that it would more closely match thermodynamic temperature, but it is also recognised that such changes would be a major disruption to the overwhelming majority of users for whom ITS-90 has been, and likely will be for the foreseeable future, a perfectly acceptable standard. However, there are other changes under consideration that will have an effect on thermometry. The first is the planned redefinition of the kelvin in terms of the Boltzmann constant. The current definition is straightforward to realise: a triple point of water cell can be purchased and after a little tuition a user can be measuring 0.01 °C. The situation will become more complex once the definition of the kelvin effectively becomes "do an experiment to determine the Boltzmann constant and choose the temperature unit to be such that k_B is equal to the specified defined value." Clearly such a definition is much more open than the situation that exists at the moment, but some guidance will be required to ensure that suitable methods are used. The situation will be similar to that when the metre stopped being defined as the length of a particular platinum-iridium rod and instead was defined by fixing the speed of light *in vacuo* with zero uncertainty. That change led to the adoption of a guide to practical realisation, or '*mise en pratique*' for the definition of the metre. A *mise en pratique* (MeP) is an updateable document that helps users make an appropriate choice of method to realise the unit. The *mise en pratique* for the definition of the kelvin (MeP-K) was adopted in 2006 in recognition that additional guidance would be required after the anticipated kelvin redefinition. At that time it just consisted of the definition of the kelvin as a fixed fraction of the thermodynamic temperature of the triple point of water, the text of the ITS-90 and the text of the Provisional Low Temperature Scale of 2000 (PLTS-2000). It has now had added documents dealing with the isotopic composition of some of the materials specified in the ITS-90 and listing the current best estimates and uncertainties for the difference between thermodynamic temperature and ITS-90. An advantage of a MeP is that it can readily be extended by adding new documents without altering what is already in place. The intention for thermometry is that ITS-90 will continue as a scale that satisfies the requirements of the majority of users. There will be another section that describes primary methods for realising thermodynamic temperature. The methods will be those that can achieve the best uncertainties. At higher temperatures this will include spectral radiometry measurements.

A second change has been the development of practical high temperature fixed point references. This has led to the expectation that the MeP-K will have a section on "indirect approximations" [7]. If we consider the recent work on determining the Boltzmann constant with low-enough uncertainty for it to be used as the definition of the kelvin, one way to do this was to measure the speed of sound in a gas held at the temperature of the triple point of water. The equation can be written exactly, and all the parameters required, such as the dimensions of the acoustic chamber, can be measured or calculated. The uncertainties in the necessary measurements were so low that the final uncertainty in k_B was at the level of 1 ppm, which combined with results from other laboratories is considered sufficient accuracy for the kelvin to be defined by giving k_B a fixed value with zero uncertainty. Of course, once k_B is fixed the apparatus previously used to measure the constant becomes a primary thermometer. The speed of sound and k_B allow some arbitrary temperature of the gas-filled cavity to be calculated from the equation of state and this can be used to calibrate sensors that are in thermal equilibrium with the apparatus.

In contrast, for indirect approximations, and as for the formal approximation of ITS-90, rather than measuring all the necessary parameters of the equation of state a series of measurements of signals at known temperatures are made and an interpolation equation can then be used to determine the signal-temperature relationship at intermediate temperatures. The first addition to the indirect approximations are planned to be the thermodynamic temperatures and uncertainties of high temperature fixed points of cobalt-carbon, platinum-carbon and rhenium-carbon eutectic alloys. An interpolation function based on the Planck function will be specified. Such a scheme will not be a primary method, nor will it be a "formal approximation" like the ITS-90 where the fixed points have defined values with zero uncertainty, but it will provide a thermodynamic based temperature scale that will be "traceable to the SI."

CHAPTER 2

Fundamental laws

2.1 INTRODUCTION

In the first chapter we have considered the basic concepts which underlie the measurement of physical quantities, and of temperature in particular. We now direct our attention specifically to the field of radiation thermometry, the estimation of the temperature of an object from the characteristics of the thermal radiation which it emits.

It has of course been known from the earliest times that objects emit visible electromagnetic radiation, or light, when they become very hot[1]. Not only does the intensity or brightness of the radiation increase rapidly with increasing temperature, but the colour perceived changes from a dull red through orange and yellow to an intense white for the highest temperatures readily achieved in practice. Both properties may be employed to estimate and control the temperature, a requirement even in pre-history for the smelting of metals and the firing of pottery. Those early workers made use of the natural radiation thermometer formed by the human eye and its associated signal conditioning system, the brain. While this 'instrument' is rather unstable and incapable of a high absolute accuracy, the eye is capable of resolving small differences, corresponding to temperature changes of a few degrees, in the brightness or colour of adjacent objects. The skilled foundryman in older times therefore worked from the appearance of a furnace, often using as a reference a piece of coloured material – orange peel is often quoted as an example.

'Optical' or 'visual' radiation thermometers, that is, those using the eye as a radiation detector, are still occasionally used in laboratories and factories, and usually contain a calibrated filament lamp as a reference source. The main disadvantage with these instruments for many applications is that they require an experienced and skilled operator to obtain the best results. The modern tendency is to prefer instruments designed around photoelectric detectors. The reasons for this will become apparent through this book, but they include greater reliability, a choice of

---

[1]Strictly, once the concept of light as some influence that travelled from the object to the eye had been established, rather than the early theory of sight being due to some emanation from the eye.

operating temperature ranges, and the ability to measure continuously with lower operating costs.

With the discovery of techniques for the detection of infrared radiation, it became apparent that even quite cool objects lose energy in this manner. These observations ultimately led to the idea that all bodies, whatever their temperatures, emit electromagnetic radiation as a result of the motion of their constituent particles. Thermal radiation, as it is generally known, covers an extraordinarily wide wavelength range, from the ultraviolet up into the microwave and radio frequency bands. In this book we shall limit discussion to wavelengths between about 400 nm and 20 μm, apart from those cases where the total energy radiated is being considered, when the upper limit may be extended to wavelengths of a few hundred micrometres.

It was soon apparent that if accurate and reliable thermometry was to be achieved, a better theoretical understanding of the laws of thermal radiation was required. It eventually turned out that two topics were fundamental to the understanding and practice of radiation thermometry – the concept of a black-body radiator, and the Planck radiation law. The first provides a reference against which the characteristics of real surfaces may be established, and which moreover may be used in the theoretical calculation of the underlying laws of emission and absorption of radiation. Planck's law, which gives the spectral distribution of the intensity of the radiation from a black-body, is fundamental to the calculation of the performance of any radiation thermometer, whatever the wavelength region and range involved in its operation. Because of their importance, we shall spend some time in this chapter discussing the properties of black-bodies and the development and derivation of Planck's law. The properties of real surfaces, on the other hand, will form the subject of the next chapter.

2.2   THE CONCEPT OF A BLACK-BODY RADIATOR

It is possible to derive many of the essential relationships required in radiation thermometry from simple arguments based upon the laws of thermodynamics and the assumption of geometrical optics, that is, that thermal radiation is composed of rays, which in a uniform isotropic medium travel in straight lines. In this case, the wave properties of electromagnetic radiation are neglected or treated as a small correction to be applied to the relationships derived from geometrical optics. Also, it will usually be assumed that the radiation is unpolarised; this is not necessarily the case, and the possible effects of partially polarised radiation upon the performance and accuracy of different radiation thermometers will be discussed where relevant.

If we isolate a beam of thermal radiation, and re-direct it onto the surface of a solid or liquid, a variety of effects may be observed, some of which are shown in Figure 2.1. First, part of the beam may be reflected at the surface. Specular or regular reflection describes that part of the beam which appears at the same angle to the normal to the surface as the incident beam, but on the other side of the normal and in the same plane. As the name suggests (*speculum* is Latin for

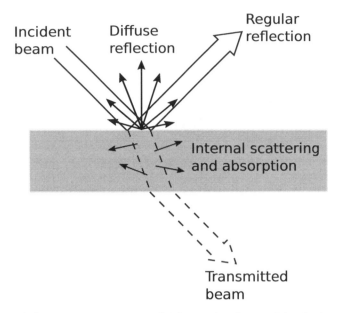

Figure 2.1: Some of the various processes which may be observed in the interaction of a light beam with solid or liquid materials.

'mirror'), this behaviour is typical of smooth, shiny surfaces. With rough surfaces the beam may be reflected at all angles from the incident face, and it is then said to be diffusely reflected. It is also possible for a proportion of the reflected radiation to appear along the line of the incident beam, but reversed in direction. If the proportion is significantly higher than that expected from the angular distribution of the diffusely reflected radiation, the phenomenon is known as retro-reflection. It is rarely an important effect in radiation thermometry, but its existence should be recognised, for example in the design of high emissivity black-body cavities.

Not all of the incident radiation will be reflected at the surface of the slab. The absorption coefficient σ, the fraction of the beam absorbed per unit length, in most materials is very large and that part of the radiation entering the slab is absorbed within a distance equal to a few wavelengths. These materials are said to be opaque to the radiation in question. In some materials, however, σ is small at some wavelengths, and a significant fraction of the radiation then appears at the rear surface. If, for a parallel sided slab, the transmitted beam travels in the same direction as the incident beam, although usually displaced in position, the material is said to be transparent. On the other hand, the beam may be scattered within the slab, and the transmitted radiation is diffuse, that is, it is spread over a wide angular range. Such materials are described as translucent. The situation is clearly more complex for these materials than for opaque solids. There are other effects not shown in Figure 2.1, for example total internal reflection within the material, and a diffuse output beam produced by scattering at the exit surface. But whatever effects are present, the law of conservation of energy dictates that the sum of the reflected, absorbed or transmitted fractions of incident beam must be unity.

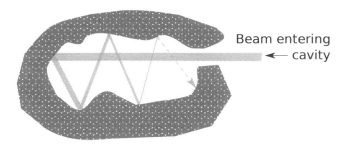

Figure 2.2: Absorbing cavity as an approximation to a black-body.

In the study of the relationships between radiative quantities it is convenient to introduce two idealised surfaces. The first is a perfect reflector, that is, a surface which completely reflects all of the radiation incident upon it. All materials which absorb thermal energy are in thermal contact with their environment. An object with a perfectly reflecting surface, however, may be totally isolated thermally from its surroundings. By placing it in a vacuum and supporting it with a thermal insulator, conductive and convective heat transfer may be eliminated. It follows that the object cannot emit thermal radiation (that is, its emissivity is zero), as otherwise its temperature would fall, eventually to absolute zero. In practice, the best mirror surfaces have reflectance around 99.99 %, over a moderate wavelength range.

The second ideal surface is one which completely absorbs all incident radiation, with none being reflected or transmitted, and which for obvious reasons is known as a black-body. Although in principle we might require that this characteristic should be shown at all wavelengths, it is sufficient for many purposes that it applies only over the limited range of current interest. Real surfaces are, after all, rather variable in this respect. Alumina, for example, appears quite white at visible wavelengths, that is, it reflects them rather well, whereas it is an effective absorber of infrared radiation, particularly when hot. In the latter case, it may be considered to be reasonably 'black'.

In fact no real surface is a particularly close approximation to a black-body, even over a limited wavelength range. The best absorb around 99 % of the incident radiation. However, a much better approximation to the ideal is provided by a cavity (see Figure 2.2) whose walls are at a uniform temperature and having a small aperture through which radiation may enter and exit. Clearly, if the cavity is appropriately designed and the material of the walls absorbs radiation even moderately well, a beam entering through the aperture will undergo multiple reflections inside and lose energy at each reflection, so that very little will eventually be re-emitted. It will be shown that an object which absorbs all incident radiation also emits with the maximum intensity allowed at that temperature, that is, it has the maximum emissivity of unity.

2.2.1 The radiation quantities

The thermal radiation field inside a closed cavity is very closely related to the amount of radiation emitted by a black-body surface. Before we derive this relationship, and the properties of the field, it is necessary to define more exactly the various radiation quantities which will be needed here and later in the book. They include:

- The total radiant energy, Q, in a particular system, for example in a black-body cavity. The energy is measured in joules, which has the symbol 'J'.

- The radiant energy density, u, is the energy per unit volume, has units $J\,m^{-3}$ and is given by

$$u = \frac{dQ}{dV}$$

- The radiant flux, Φ, is the rate of radiant energy transfer from one region to another and is given in units of joules per second $(J\,s^{-1})$ or equivalently in watts (W).

- The radiant exitance, M, is the total flux emitted in all directions per unit area of radiating surface

$$M = \frac{d\Phi}{dA}$$

  The flux per unit area received by a surface from all external directions is a similar quantity, known as the irradiance, E. Both are expressed in units of watts per square metre $(W\,m^{-2})$.

- The radiance, L, is the flux emitted per unit solid angle in a direction θ to the normal to the surface, per unit area projected along this direction,

$$L = \frac{d^2\Phi}{dA d\Omega} \cos\theta$$

  The units are $W\,m^{-2}\,sr^{-1}$, where 'sr' stands for the unit of solid angle Ω, the steradian.

- The radiant intensity, I, is the flux emitted into a unit solid angle in a given direction and is expressed in watts per steradian, $(W\,sr^{-1})$

$$I = \frac{d\Phi}{d\Omega}$$

  These quantities are defined for the total radiation emitted by, or incident upon, a surface over all wavelengths. A similar set of quantities may be defined for radiation at a particular wavelength λ. If the total radiation quantity is represented by X, the corresponding spectral quantity X_λ is defined as

$$X_\lambda = \frac{dX}{d\lambda}$$

that is, it is the density of the quantity per unit wavelength interval. The corresponding units are the same as those for X with the addition of the inverse of wavelength interval. This is conveniently expressed in terms of nanometres or micrometres, so that the unit of spectral radiance, I_λ, for example, might be given as $\mathrm{W\,m^{-2}\,sr^{-1}\,nm^{-1}}$. It is convenient to use a different sub-multiple of length for the wavelength interval from that for the emitting area, to draw attention to the fact that they are essentially different quantities (but see the discussion in §2.10 on this point). Either nanometres or micrometres may be employed, but it is advisable to adopt one and employ it consistently in order to avoid numerical errors.

The spectral distribution of a radiation quantity X is normally given as a function of wavelength; less commonly, at least in radiation thermometry, the frequency, ν, may be preferred:

$$X_\nu = \frac{dX}{d\nu}$$

Since the product of the wavelength and the frequency of electromagnetic radiation in free space is equal to the speed of light c, X_ν may be calculated from

$$X_\nu = \left(\frac{c}{\nu^2}\right) X_\lambda \qquad (2.1)$$

It should be stressed that λ is by convention the wavelength *in vacuo*. Tables of spectral lines are often given in terms of the wavelength in air, and the values should be corrected if used in the calibration of radiation thermometers.

There are indeed many other radiative quantities which may be found in articles on radiometry and photometry. We shall use them to a very limited extent in this book, as they are often a source of confusion rather than of assistance in the comprehension of the subject. This is due partly to the sheer number of quantities involved, and partly from the fact that many are derived using specific assumptions about the characteristics of the emitting surface or of the behaviour of the radiation thermometer. For example, quantities may be derived under the assumption that the surface behaves as a 'grey-body', that is, that its spectral emissivity is independent of wavelength. While this may be adequate for measurements of moderate accuracy, it is rarely valid for real surfaces, and the consequent error may be important in critical applications. It is preferable to start from exact, if limited, equations so that the assumptions involved in their application may be explicitly stated and their validity examined.

2.2.2 Properties of the cavity radiation field

We now consider two black-body surfaces A and B facing each other, as shown in Figure 2.3a, their rear surfaces being perfect reflectors so that they are isolated from the environment. It is assumed that they are sufficiently large in comparison to their separation that energy losses at the periphery may be neglected. The thermal radiation emitted by one surface and totally absorbed by the other constitutes a flow of heat. From the laws of thermodynamics we know that the heat flows must balance when A and B are in thermal equilibrium, that is, at the same temperature. It follows that the thermal energy radiated by a black-body does not depend upon the nature of the surface but only upon its temperature, that is, the exitance may be written in the form $M = F(T)$. If A is at a higher temperature than B, however, there must be an overall heat flow from A to B, so that the exitance must increase monotonically with temperature.

(a) Total exitance   (b) Spectral exitance

Figure 2.3: Total and spectral energy transfer between two black-body surfaces.

By placing an ideal filter between A and B which transmits at a single wavelength λ_0 and reflects all other radiation, as shown in Figure 2.3b, it may be concluded that a similar relationship holds for the spectral distribution, that the spectral exitance depends only upon the wavelength and temperature,

$$M_\lambda = F(\lambda, T)$$

The relationship between the energy density u of the radiation field inside a closed cavity whose walls are at a uniform temperature T and the radiance and exitance of a black-body surface at the same temperature will now be developed. We introduce into the cavity a thin disc, one side of which is a perfect reflector, the other a black-body surface, and allow the system to come to equilibrium. As the disc is in thermal contact with the walls of the cavity through the radiation field, its temperature will be the same as that of the walls. At equilibrium, the rate at which the disc absorbs energy from the radiation field will be equal to the rate of thermal emission.

In a spherical polar coordinate system with its origin at the centre of the disc, as shown in Figure 2.4, the dependence of the local radiation density upon direction may be written generally as $u(r, \theta, \phi)$. θ is known as the zenith angle or colatitude and measures the deviation of the vector r from the normal ON to the disc. The azimuth ϕ is the deviation of the projection of r on the plane of the disc from some arbitrary axis OA. The elemental volume dV at (r, θ, ϕ) is given by

$$dV = r^2 \sin\theta \, dr \, d\theta \, d\phi$$

The energy content of this volume is therefore

$$dQ = u(r, \theta, \phi) r^2 \sin\theta \, dr \, d\theta \, d\phi$$

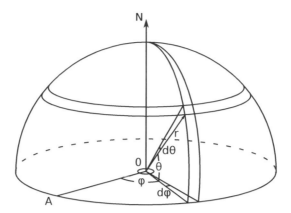

Figure 2.4: Hemispherical polar coordinate system for the thermal radiation emitted and absorbed by a flat disc placed at an arbitrary origin.

The fraction f of dQ which will be absorbed by the disc of area dA at O is that moving within the solid angle subtended at (r, θ, ϕ) by the disc,

$$f = \frac{dA \cos \theta}{4\pi r^2}$$

where $dA \cos \theta$ is the area of the disc projected along r. The total energy absorbed by the disc within a time dt must originate within a hemisphere of radius $c.dt$, where c is the speed of light in the medium filling the cavity. From the definitions given in §2.2.1 this quantity will be seen to be related to the irradiance E, that is,

$$E \, dA \, dt = \frac{dA}{4\pi} \int_0^{2\pi} \int_0^{\pi/2} \int_0^{cdt} u(r, \theta, \phi) \, dr \, \sin \theta \, \cos \theta \, d\theta \, d\phi$$

At equilibrium, the radiation field $u(r, \theta, \phi)$ is stable, at least when averaged over moderate intervals of time to smooth out the quantum fluctuations. Moreover, the rate of energy gain by the disc at a given temperature is constant, and does not depend upon the time interval dt selected. It follows that $u(r, \theta, \phi)$ must be independent of r, and the inner integral may be calculated trivially to give

$$E \, dA \, dt = \frac{c \, dA \, dt}{4\pi} \int_0^{2\pi} \int_0^{\pi/2} u(\theta, \phi) \, \sin \theta \, \cos \theta \, d\theta \, d\phi \tag{2.2}$$

If the disk and its associated coordinate system are rotated about the origin, the equation is modified to

$$E \, dA \, dt = \frac{c \, dA \, dt}{4\pi} \int_0^{2\pi} \int_0^{\pi/2} u(\theta + \Delta\theta, \phi + \Delta\phi) \, \sin \theta \, \cos \theta \, d\theta \, d\phi$$

However, the total energy radiated by the disk is independent of orientation, and hence the energy absorbed must be constant. As the changes $\Delta\theta$ and $\Delta\phi$ are arbitrary and independent, the radiation field must therefore be independent of the direction, that is, it is isotropic. Similar arguments apply when the disk is moved within the cavity, and we may also conclude that it is homogeneous. It is important to note that these conclusions are independent of the geometry of the cavity. Evaluating the integrals in (2.2) under isotropic conditions leads to an expression for the total irradiance E incident on the disk, and hence to its exitance M:

$$E = M = (c/4)u \qquad (2.3)$$

Not only is the total energy absorbed in unit time equal to that radiated, but in order to maintain the isotropic nature of the field, the same must be true for the energy flow in any given direction (θ, ϕ). From the definition of the radiance L given above, the energy radiated in a time dt to the volume element dV from the disk is given by

$$L \, dA \, \cos\theta \, d\Omega \, dt = L \, dA \, \sin\theta \, \cos\theta \, d\theta \, d\phi \, dt$$

From the derivation of (2.2), the energy absorbed from this direction in time dt is

$$(c/4\pi) \, dA \, dt \, u \, \sin\theta \, \cos\theta \, d\theta \, d\phi$$

from which it follows that

$$L = \left(\frac{c}{4\pi}\right) u \qquad (2.4)$$

and hence $\qquad M = \pi L \qquad (2.5)$

Although these relationships have been derived for a plane disc inside a closed cavity, they also apply to the inner wall of the cavity. This is simply proved by moving the disc arbitrarily close to the cavity wall. Moreover, if a small aperture is created in the cavity wall, the equations above apply also to the thermal radiation emitted through it, as long as the energy loss is not so great that it significantly perturbs the energy density. In practice, the energy loss cannot be made so small that the aperture acts as a perfect black-body. But the change in the effective emissivity from unity may be made very small, less than 1×10^{-4} in special cases, and may moreover be estimated from the geometry and composition of the cavity walls.

Since we have shown that u is homogeneous, it follows that the radiance L is independent of direction for a black-body surface. Surfaces which approximate this condition are known as diffuse emitters. It corresponds to the property that the apparent brightness of the surface is independent of the viewing angle. Paper, for example, possesses this characteristic when uniformly illuminated. The radiant intensity I, the flux emitted per unit solid angle in a particular direction from a radiating surface of area dA, may be written simply as

$$I = \int_A L \, dA \, \cos\theta$$

which for a perfectly diffuse emitter integrates trivially to

$$I = I_0 \cos\theta \qquad (2.6)$$

where I_0 is the maximum intensity observed normal to the surface. This relationship is known as Lambert's law. The radiant intensity when viewed away from the normal falls away due entirely to the decrease in the projected area of the source. This law should be applied with caution as few real surfaces behave as perfectly diffuse emitters, particularly at large angles from the normal.

Finally, it should be noted that all the relationships derived above for total radiation quantities also apply equally to their spectral derivatives.

2.3   THE STEFAN-BOLTZMANN LAW

In the previous section we showed that the total radiant energy, that is, integrated over all wavelengths, from a black-body source is a function of its temperature only. On the basis of earlier measurements, Stefan suggested in 1879 that it should follow a fourth power law. Although the experimental evidence for this was sparse, this conjecture was soon supported by a simple theoretical argument derived by Boltzmann in 1884 which will be summarised here.

First, it may be noted that since the density u of the radiant energy in a black-body cavity is a function of the temperature alone, the only non-zero differential is du/dT. It had already been shown that it was possible to define a radiation pressure p given for isotropic radiation by

$$p = \frac{u}{3} \qquad (2.7)$$

Boltzmann considered the thermal radiation existing in a cylinder of volume V closed with a reflecting piston. If the volume expands reversibly by dV, by allowing the piston to move, the heat input required to keep the temperature T constant is given by the first law of thermodynamics,

$$dQ = dU + p\,dV$$

where dU is the change in the total radiant energy in the cylinder and pdV the work done on the piston. Since $U = uV$, after substituting for p from (2.7), this becomes

$$dQ = V\,du + \frac{4}{3}u\,dV$$

The change dS in entropy is given by

$$\begin{aligned} dS &= \frac{dQ}{T} \\ &= \frac{V}{T}du + \frac{4}{3}u\frac{dV}{T} \\ &= \frac{V}{T}\frac{du}{dT}dT + \frac{4}{3}u\frac{dV}{T} \end{aligned}$$

and hence

$$\frac{\partial S}{\partial T} = \frac{V}{T}\frac{du}{dT} \qquad : \qquad \frac{\partial S}{\partial V} = \frac{4}{3}\frac{u}{T} \tag{2.8}$$

We may express the fact that for a reversible process the entropy is a function solely of the state of the system by noting that it is an exact differential, and hence

$$\frac{\partial^2 S}{\partial V \partial T} = \frac{\partial^2 S}{\partial T \partial V}$$

Deriving these terms from (2.8) we obtain

$$\frac{1}{T}\frac{du}{dT} = \frac{4}{3}\left(\frac{1}{T}\frac{du}{dT}\right) - \frac{4u}{3T^2}$$

from which it follows that

$$\frac{du}{dT} = 4\frac{u}{T}$$

and integrating

$$u = \alpha T^4$$

The constant of integration must be zero as the thermal energy density disappears at absolute zero. Applying (2.3)

$$M = \sigma T^4$$

where α and σ are related constants, the latter being known as the Stefan-Boltzmann constant. The derivation of this equation was of immediate value in that it provided a method of estimating temperatures above the range of conventional thermometry at the time. The technique suffered from a number of practical difficulties, principally that of ensuring that the bolometer detectors employed responded equally to all wavelengths radiated.

2.4   THE DEVELOPMENT OF PLANCK'S LAW

Well before the form of the Stefan-Boltzmann law had been established, there had been considerable interest in the spectral distribution of the thermal radiation from hot surfaces. In 1847 John William Draper had found that the thermal spectrum was smooth and continuous, being free from the bands and spectral lines found in many other sources of visible radiation. The intensity showed a broad maximum, falling away quite rapidly at short wavelengths, but possessing a pronounced tail at long wavelengths. The wavelength of maximum intensity was found to depend upon the temperature of the surface. It was eventually concluded from experimental measurements and on theoretical grounds that this dependence could be summarised by an equation now known as Wien's displacement law

$$\lambda_m T = c_3 \tag{2.9}$$

where c_3 was a constant, sometimes known as the third radiation constant and approximately equal to $3000\,\mu\text{m}\,\text{K}$.

The displacement law was also employed to estimate the temperature of hot surfaces, including that of the sun. However, it is difficult to locate precisely the position of a maximum when the distribution is broad, and the result is sensitive to the presence of wavelength-dependent errors in the measurement of the spectral intensity. It became clear that for precise thermometry it would be necessary to establish the equation describing the spectral distribution of thermal radiation, and several groups, mainly situated in Germany, set to work to tackle the problem.

The first major step was taken by Wien, who in 1893 extended Boltzmann's treatment of thermal radiation in a cylindrical cavity with a reflecting piston to consider the spectral distribution. Although he was unable to derive a complete expression for the spectral distribution of black-body radiation, he developed scaling laws, including a general form of the displacement law (2.9). These, together with the necessary condition that the integral of M over all wavelengths must lead to the Stefan-Boltzmann law, led to the conclusion that the distribution had to be of the general form

$$M(\lambda, T) = \lambda^{-5} g(\lambda T) \tag{2.10}$$

where $g(\lambda T)$ was an unknown function of the product λT.

Wien proposed, partly on questionable theoretical grounds, that the correct form for $g(\lambda T)$ was such that

$$M(\lambda, T) = c_1 \lambda^{-5} \exp(-c_2/\lambda T) \tag{2.11}$$

where c_1 and c_2 were known as the first and second radiation constants, respectively. This equation, now known as the Wien distribution law or the Wien approximation, fitted the experimental data then available rather well. Most measurements at that time covered wavelengths around the maximum in the spectral distribution. As techniques improved and the wavelength range was extended into the infrared, small deviations from the Wien equation (2.11) were found. To correct for these, Planck made what appeared to be a minor modification,

$$M(\lambda, T) = c_1 \lambda^{-5} \left[\exp\left(\frac{c_2}{\lambda T} \right) - 1 \right]^{-1} \tag{2.12}$$

At first the basis for this was largely empirical, but within a few months he was able to provide a theoretical derivation. It did, however, require a slightly odd assumption, and this was the first indication that the postulates of classical physics were inadequate to explain the spectral distribution of black-body radiation.

The arguments presented here do not follow Planck's exactly; the treatment has been simplified and later developments incorporated in order to illustrate the relationships with other fundamental concepts. We start from a modified form of one of Planck's early equations,

$$u(\nu, T) = D(\nu) E(\nu, T)$$

which simply states that the radiation density is the product of the density of oscillators, or modes of oscillation, in a given frequency interval and their mean energy. Planck initially spoke only of resonators in the cavity itself, although later writers assumed that his equations really applied to the atomic or molecular oscillators in the walls of the cavity, since the cavity itself could be considered to be completely empty of material. Nowadays, the resonators are again taken to correspond to modes of oscillation of the radiation field within the cavity, each mode as we shall see behaving as a simple harmonic oscillator. It was first pointed out by Rayleigh and Jeans that $D(\nu)$, the frequency distribution of the modes of the radiation field within the cavity, could be calculated directly from Maxwell's equations.

As we have already demonstrated that the radiation field for closed cavities at a given temperature is independent of their geometry or the properties of the walls, we are free to assume a convenient form for our calculations. In fact, we shall take a rectangular parallelepiped of sides L_x, L_y, L_z, where these are assumed to be large in comparison to the wavelengths of the radiation under consideration. The radiation field in the cavity may be represented by the vector potential $A(\boldsymbol{r}, t)$, where the functional dependence on the position vector $\boldsymbol{r}$ and on the time t is wave-like, that is, it may be taken as a set of sinusoids. If the walls of the cavity are good reflectors of electromagnetic radiation, boundary conditions of the form

$$\exp i k_x x = \exp[i k_x (x + L_x)]$$

apply, as the solutions are standing waves with nodes at the walls. These conditions may also be written in the form

$$k_x = \pi N_x / L_x$$

where N_x, N_y and N_z are positive integers (negative values would change the directions of the travelling waves, but leave the standing wave solution unaltered). The components k_x, k_y and k_z define the wave vector

$$\boldsymbol{k}_p = 2\pi (k_x \boldsymbol{i} + k_y \boldsymbol{j} + k_z \boldsymbol{k})$$

where $\boldsymbol{i}$, $\boldsymbol{j}$ and $\boldsymbol{k}$ are the unit vectors in the x-, y- and z-directions, respectively. The boundary conditions therefore lead to a set of discrete modes characterised principally by the wave vector $\boldsymbol{k}_p$ (not to be confused with the unit vector $\boldsymbol{k}$) or alternatively by the mode numbers (N_x, N_y, N_z). In addition, we require a real unit vector $\boldsymbol{p}_{k\lambda}$ to describe the polarisation state of each individual mode. A general form for the vector potential may be given as a linear superposition of the terms for the individual modes:

$$A(\boldsymbol{r}, t) = \sum_{k,\lambda} \{ A_{k\lambda} \exp[i(\boldsymbol{k}_p \cdot \boldsymbol{r} - 2\pi \nu_k t)] + A_{k\lambda}^* \exp[i(\boldsymbol{k}_p \cdot \boldsymbol{r} - 2\pi \nu_k t)] \} \boldsymbol{p}_{k\lambda} \quad (2.13)$$

where $A_{k\lambda}$ and $A_{k\lambda}^*$ are the amplitude and its complex conjugate of a mode characterised by the wave and polarisation vectors $\boldsymbol{k}_p$ and $\boldsymbol{p}_{k\lambda}$.

Excluding for the moment the different polarisation states, the mode density may be calculated when the number of modes is very large, that is, it may be treated as continuous, from

$$
\begin{aligned}
dN &= dN_x \, dN_y \, dN_z \\
&= (L_x \, L_y \, L_z \, / \pi^3) \, dk_x \, dk_y \, dk_z \\
&= (V/\pi^3) \, dk_x \, dk_y \, dk_z
\end{aligned}
$$

The term $dk_x \, dk_y \, dk_z$ is an element of volume in k-space. From (2.13) it may be seen that if the wave velocity in the medium is c the magnitude k of the wave vector $\boldsymbol{k}_p$ is

$$
k = 2\pi\nu/c \tag{2.14}
$$

It follows that a given frequency corresponds to a fixed value of k, that is, to points on the surface of a sphere of radius k in k-space. However, as has been noted, only the positive values of k_x, k_y and k_z are significant in the case of standing waves, and we may limit our attention to the positive octant of the sphere. To obtain the number of modes in a given frequency interval $d\nu$, we note that the element between the octants of radii k and $(k + dk)$ has volume $(\pi/2)k^2 dk$, and hence

$$
dN = (V/\pi^3)(\pi/2)k^2 dk
$$

which, after substituting for k from (2.14), becomes

$$
dN = \frac{4\pi V}{c^3} \nu^2 d\nu
$$

To allow for the different possible polarisation of the modes, we note that the polarisation state vector $\boldsymbol{p}_{k\lambda}$ has two components $\boldsymbol{p}_{k\lambda}$ and $\boldsymbol{p}_{k\lambda'}$ satisfying the condition

$$
\boldsymbol{p}_{k\lambda}\boldsymbol{p}_{k\lambda} - \delta_{\lambda\lambda'}
$$

It might be expected from the general rotational properties of a vector field that we could associate with it an intrinsic spin $S = 1$, implying that there would be three eigenvectors of values -1, 0 and 1. However, the Coulomb gauge

$$
\nabla \cdot \boldsymbol{A} = 0
$$

ensures the transversatility of the electromagnetic field, that is, that the electrical and magnetic vectors $\boldsymbol{E}$ and $\boldsymbol{B}$ are transverse to $\boldsymbol{k}_p$, so that $\boldsymbol{p}_{k\lambda}$, $\boldsymbol{p}_{k\lambda'}$ and $\boldsymbol{k}_p$ form a mutually orthogonal set of unit vectors. The spin eigenvector in the direction of propagation is therefore forbidden, and there are indeed only two polarisation states. We must therefore multiply the expression for dN above by two to allow for the fact that each mode has this number of orthogonal states of polarisation. The final result for the mode density is, therefore,

$$
D(\nu) = \frac{1}{V} \frac{dN}{d\nu} = \frac{8\pi\nu^2}{c^3} \tag{2.15}
$$

2.4.1 Boltzmann statistics

We now consider the calculation of the mean energy per mode $E(\nu, T)$. The method developed by Planck was based upon the statistical methods initiated by Boltzmann around 1872, and applied mainly to problems in ideal gas behaviour. The aim was to calculate the number W of different ways a total of N distinguishable particles could be distributed among the accessible energy levels of the system. W was related to the entropy S of the system by

$$S = k \ln W \tag{2.16}$$

where k is the Boltzmann constant. An essential element of the method was the postulate that the most probable distribution of particles was that which gives the greatest value for W, that is, the arrangement with the maximum entropy. For convenience we may express this condition in the form

$$d \ln W = 0$$

We suppose that N_i particles enter the levels characterized by energy E_i and degeneracy g_i. The number of ways that N objects may be divided into groups of N_i is

$$W_i = \frac{N!}{\prod_i N_i!}$$

Within each group the N_i particles may be arranged in different ways amongst the g_i energy levels. If it is assumed that there are no restrictions upon the number of particles occupying a particular energy level, the total number of permutations W is given by

$$W = N! \prod_i \frac{g_i^{N_i}}{N_i!} \tag{2.17}$$

For convenience, we adopt a simple form of Stirling's approximation

$$\ln N! = N \ln N - N$$

The neglect of the multiplying constants in the more complete expression may be justified on the grounds that they are small in comparison with the terms retained, and that we are in any case concerned with relative probabilities as we are seeking the maximum value of W. Applying this approximation to the logarithm of (2.17)

$$\ln W = \ln N! + \sum (N_i \ln g_i - \ln N_i!)$$

yields

$$\ln W = N \ln N - N + \sum (N_i \ln g_i - N_i \ln N_i + N_i) \tag{2.18}$$

Since $N = \sum N_i$, these two terms cancel on the right-hand side. Differentiating the result with respect to N_i gives

$$d \ln W = \sum [\ln(g_i/N_i)]dN_i \tag{2.19}$$

We also need to include the conditions that the total number of particles N and the total energy E are constant. These may be introduced in the form under the conditions

$$\sum dN_i = 0 : \sum E_i dN_i = 0$$

Combining them with (2.19) using the technique of Lagrangian multipliers, the overall equation becomes

$$\sum [\ln(g_i/N_i) - 1 + \alpha + \beta E_i]dN_i = 0$$

As the dN_i are both arbitrary and independent, this equation can only be satisfied if each term within the summation is separately zero, that is,

$$\ln g_i/N_i = 1 - \alpha - \beta E_i$$

Substituting this back into (2.18),

$$\ln W = N \ln N + \sum N_i(1 - \alpha - \beta E_i)$$
$$= N \ln N + N(1 - \alpha) - \beta E$$

With (2.16), this gives an expression for the entropy S which may be related to temperature T through

$$\frac{dS}{dE} = \frac{1}{T}$$

Carrying out the differentiation enables the constant β to be determined:

$$\beta = -1/kT$$

and hence

$$N_i \propto g_i \exp(-E_i/kT) \tag{2.20}$$

which is a form of the Maxwell-Boltzmann distribution. The constant of proportionality, which includes the constant α, in any given case is obtained by summing over all the energy levels. It may be noted that the calculation has made few assumptions about the nature of the system, and it is as a result very general in its application. If the energy levels are closely packed, so that they may be treated as a continuous distribution, and if the energy depends quadratically upon either the coordinates or the momentum of the particle, then the equation may be integrated to give

$$E = kT/2 \tag{2.21}$$

that is, the mean energy for each particle or mode obeying the quadratic requirement is $kT/2$. The total energy is therefore made up of a number of equal contributions from each relevant mode of motion; this is known as the principle of equipartition of energy. In the case of the radiation field inside a black-body cavity, combining this result with the density of modes obtained in (2.15) gives the Rayleigh-Jeans law for the spectral distribution:

$$u(\nu, T) = 8\pi\nu^3 kT/c^3 \tag{2.22}$$

Clearly this does not predict a maximum in the spectral distribution at any frequency or wavelength, and must therefore be incorrect in a general sense. Although it is frequently stated that it does predict the spectral intensity at low frequencies (long wavelengths) and high temperatures, for the purposes of radiation thermometry the agreement, as shown in Figure 2.5, is rather poor. In fact, the main importance of the Rayleigh-Jeans law was that it demonstrated that classical physics at the turn of the century could not correctly calculate the spectral distribution of black-body radiation, and that some additional postulates were required.

2.4.2 Quantum statistics

We now consider the modification which Planck found necessary in order to provide a theoretical derivation of his empirical equation. Following the Boltzmann technique outlined in the previous section, the distribution of modes (or resonators in the Planck formulation) is divided into groups, each covering a small frequency interval. A given group may contain G_i modes, and correspond to an energy E_i. At this point Planck followed Boltzmann's methods very closely. To allow statistical methods to be applied, he divided the total energy $G_i E_i$ arbitrarily into N_i units of energy of value ϵ_i. It was implicitly assumed that these units were the same and indistinguishable, although the full significance of this step was not recognised until much later. In Boltzmann's applications of the method the assumption was not of great importance as a continuous energy distribution was restored towards the end of the calculation by allowing the magnitude of ϵ_i to tend to zero.

Following this approach, the number of ways W_i of distributing the N_i units of energy amongst the G_i resonators is given by

$$W_i = \frac{(N_i + G_i - 1)}{N_i!(G_i - 1)}$$

and for the whole system

$$W = \prod_i W_i$$

Again, it is assumed that the most likely state is that with the greatest value of W. Combining these equations and applying the simple form of Stirling's approximation yields after differentiation with respect to the N_i

$$d \ln W = \sum_i [\ln(N_i + G_i - 1) - \ln N_i] dN_i$$

In this case, however, we cannot assume that the total number of particles is constant, as units of energy (or, as we would now say, photons) may be created by emission or destroyed by absorption processes in the cavity. At a given temperature, however, the total energy is fixed and this leads to the condition

$$\sum_i \epsilon_i dN_i = 0$$

Applying the method of Lagrangian multipliers and following the same arguments as in the previous section, we arrive at the equation

$$\ln \left[\frac{(N_i + G_i - 1)}{N_i} \right] + \beta \epsilon_i = 0$$

Again, we may show with the same arguments that the Lagrangian multiplier β is given by $(-1/kT)$. Since both N_i and G_i may be large, the term -1 in the numerator may be neglected without serious error. Rearrangement leads to the equation

$$N_i = G_i \left[\exp \left(\frac{\epsilon_i}{kT} \right) - 1 \right]^{-1} \tag{2.23}$$

Particles obeying this equation are now said to follow Bose-Einstein statistics, and are collectively called 'bosons'.

The mean energy for each resonator or mode of frequency is easily obtained from this equation, as

$$E(\nu, T) = \frac{N_i \epsilon_i}{G_i}$$
$$= \epsilon_i \left[\exp \left(\frac{\epsilon_i}{kT} \right) - 1 \right]^{-1}$$

If we now attempt to return to a continuous energy distribution by allowing ϵ_i to tend to zero, we find that $E(\nu, T)$ tends to a constant value of kT, leading only to the classical results represented by the limited Rayleigh-Jeans equation for the spectral distribution. To obtain the result he required, Planck was forced to terminate the calculation without introducing this final step. This may be justified by assuming that the resonators absorb or emit energy in discrete amounts ϵ_i, or, more stringently, that their energy is restricted to multiples of this value. By comparison with Wien's equation in the region where this is valid, at high frequencies, Planck concluded that the unit had to be proportional to the frequency, that is

$$\epsilon_i = h\nu$$

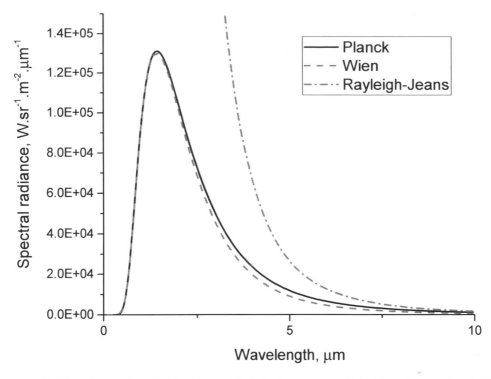

Figure 2.5: The Planck distribution for a black-body source at 2000 K *and the related Wien and Rayleigh-Jeans approximations.*

where h is a universal constant, now known appropriately as Planck's constant. Combining these expressions with the mode density (2.15) we arrive at Planck's equation in terms of frequency

$$u(\nu, T) = \frac{8\pi h\nu^3}{c^3} \left[\exp\left(\frac{h\nu}{kT}\right) - 1 \right]^{-1} \tag{2.24}$$

The spectral distribution is shown in Figure 2.5 against wavelength for a black-body at a temperature of 2000 K. It will be seen that the bulk of the radiation emitted lies in the near infrared. The curve is quite asymmetric, and tails off quite slowly at long wavelengths. The Wien approximation is very good at short wavelengths, in the region where most narrow band radiation thermometry is practised, but is always low on the long wavelength side of the peak. The Rayleigh-Jeans equation, on the other hand, is always high, and is a poor approximation everywhere in the region covered by the figure.

Thus far we have concentrated on the spectral distribution of thermal radiation; in studying the fundamental limitations of photo-detectors we need to estimate the fluctuations in the radiation field itself. From (2.23), the average number n_i of photons in a given mode of the radiation field, called the photon occupation number or mode occupancy, is given by

$$n_i = \frac{N_i}{G_i} = \left[\exp\left(\frac{\epsilon_i}{kT}\right) - 1\right]^{-1} \tag{2.25}$$

The probability distribution giving the variation in the number of photons in this mode is obtained from the Maxwell-Boltzmann distribution

$$P(n) = A\exp\left(-\frac{nh\nu}{kT}\right)$$

where A is a normalisation factor which may be calculated by summing $nP(n)$ over all n and equating the answer to the occupancy given in (2.25). The result leads to

$$P(n) = \exp\left(-\frac{nh\nu}{kT}\right)\left[\exp\left(\frac{h\nu}{kT}\right) - 1\right] \tag{2.26}$$

From this we may easily calculate the variance in n:

$$\sigma_n^2 = \sum_{n=1}^{n=\infty} n^2 P(n)$$
$$= n_i(n_i + 1) \tag{2.27}$$

When the mode frequency ν is high, n_i is small, with the result that the variance is equal to the n_i, the average number of photons in the mode. This Poisson statistical behaviour is a characteristic of particles. When ν is low and n_i is large, the variance is given by n_i^2, that is, it shows the excess noise which is characteristic of wave interactions. The point, first made by Einstein in 1906, is another indication of the inability of classical physics to account for the behaviour of photons at short wavelengths.

The radiation field at a point on a photodetector produced by a distant source does not consist of a single mode, even if it is nearly monochromatic, but may be represented by the sum of the travelling wave components from (2.13). For a chaotic thermal source, the amplitudes of the individual modes fluctuate as a result of the variation in the photon occupation numbers from the Bose-Einstein statistics. The central limit theorem indicates that the independent sums of the real and imaginary parts from the complex amplitude coefficients $A_{k\lambda}$ are random with Gaussian probability distributions. The effects of these fluctuations are discussed in §2.9.

2.5 EINSTEIN'S DERIVATION OF PLANCK'S LAW

One of the results of the development of quantum theory is that it is difficult to make positive statements about the position and even the identity of fundamental particles. It is not possible to track the motion of particles in the traditional sense, as each observation of their position alters the trajectory by virtue of the uncertainty principle. In addition, if two identical particles are placed in a box, it is impossible at a later time to say which is which. This interchangeability of particles has major effects upon the wave functions which describe their allowed characteristics of motion.

Consider two weakly interacting identical particles at positions r_1 and r_2 with respect to some convenient origin in space. Separately they may be described by eigenfunctions Ψ_1 and Ψ_2; the conventional interpretation suggests that $\Psi_1(r_1)\Psi_1^*(r_1)$ represents the probability of finding particle 1 at position r_1. The overall wave function for the two-particle system when they are in close proximity may be written as the sum of products of the form

$$\Psi = c_1\Psi_1(r_1)\Psi_2(r_2) + c_2\Psi_1(r_2)\Psi_2(r_1)$$

The coefficients c_1 and c_2 are subject to the restriction that the probability $\Psi\Psi*$ should not change when the particles are interchanged, as they are indistinguishable. This condition allows two forms for Ψ, one symmetric and one anti-symmetric with respect to the exchange of the particles:

$$\Psi_S = \frac{1}{\sqrt{2}}[\Psi_1(r_1)\Psi_2(r_2) + \Psi_1(r_2)\Psi_2(r_1)]$$

$$\Psi_A = \frac{1}{\sqrt{2}}[\Psi_1(r_1)\Psi_2(r_2) - \Psi_1(r_2)\Psi_2(r_1)] \tag{2.28}$$

This behaviour may be generalised without difficulty to systems consisting of several or many particles. If the overall wave function is anti-symmetric, the particles must differ in some way, otherwise the two terms on the right-hand side of (2.28) would be identical and cancel, reducing the eigenfunction to zero. This requirement is the basis of the Pauli exclusion principle, which applies to particles with half-integer spins. An important example of its effects lies in the structure of the electron orbitals around atoms. Each orbital can contain only one electron, and must be distinguished by different values for the three quantum numbers n, l and m, and for the spin quantum number s. The principle provided the first reasonable explanation for the existence of groups of elements with similar physical and chemical characteristics, and hence for the structure of the periodic table.

If the restriction upon the occupation of energy levels is included in a calculation of the population distribution using the methods of statistical mechanics, the particles are found to obey Fermi-Dirac statistics,

$$N_i = G_i\left[\exp\left(\frac{\epsilon_i}{kT}\right) + 1\right]^{-1} \tag{2.29}$$

Fundamental particles with half-integer spin are therefore known as fermions. They may be contrasted with those particles with integer values for the spin, photons for example, which have symmetric wave functions. There is consequently no restriction upon the occupancy of a given state and, as a result, they obey the Bose-Einstein statistics described previously.

Einstein was probably the first scientist to fully appreciate the significance of Planck's work on the spectral distribution of black-body radiation. In the years around 1906 he published a series of papers that studied the fundamental nature of electromagnetic radiation and, at a time when almost every other scientist was convinced of the wave nature of light, suggested that it could show some of the properties of particles. (It was not until much later, in 1926, that the name 'photon' was coined to emphasise this point.) In one of these papers he provided a simple derivation of Planck's law. As it is short and illustrates some basic theoretical concepts which we shall require later, it will be summarised here.

We consider atoms with discrete energy levels situated in the radiation field of density $u(\nu, T)$ within a black-body cavity, and in particular those transitions which take place as a result between the levels with energies E_i and E_j. The probability dW that an atom will absorb energy from the field and be excited from level E_i to E_j during a time dt may be written as

$$dW_{ij} = B_{ij} u(\nu, T) dt$$

where B_{ij} is an atomic constant, independent of temperature, known as the Einstein coefficient of stimulated emission when energy is liberated or, as in this case, the Einstein coefficient of absorption. Conversely, the probability of a transition from the upper level j to the lower i is given by

$$dW_{ji} = B_{ji} u(\nu, T) dt + A_{ji} dt$$

where A_{ji} is the Einstein coefficient of spontaneous emission, that is, emission which takes place independently of the radiation field.

The number of atoms in a level is given by the classical Boltzmann distribution, (2.20):

$$N_i \propto G + i \exp\left(-\frac{E_i}{kT}\right)$$

In equilibrium, the number of atoms excited into the upper level must balance those lost by the spontaneous and stimulated emission processes, that is,

$$N_i dW_{ij} = N_j dW_{ji}$$

Substituting for each factor gives

$$G_i \exp\left(-\frac{E_i}{kT}\right) B_{ij} u(\nu, T) = G_j \exp\left(-\frac{E_j}{kT}\right) [B_{ji} u(\nu, T) + A_{ji}] \qquad (2.30)$$

In the limit, at very high temperatures, the number of atoms in each level becomes the same, and the contribution from spontaneous emission becomes negligible in comparison to that from stimulated emission, as the density of black-body radiation increases monotonically with temperature. Under these conditions, the equation simplifies to

$$G_i B_{ij} = G_j B_{ji}$$

However, as each of the factors involved is an atomic constant, this equation must be valid at all temperatures. Substituting it back into (2.30) and rearranging gives

$$u(\nu, T) = \frac{A_{ji}}{B_{ji}} \tag{2.31}$$

It was noted in §2.4 that Wien had demonstrated that $M(\lambda, T)$ had to be of a particular form. Transforming (2.10) using (2.1) and (2.3) gives the equivalent form for the energy density in terms of frequency

$$u(\nu, T) = a\nu^3 F(\nu/T)$$

which will be satisfied if

$$A_{ji}/B_{ji} = a\nu^3 \qquad : \qquad E_j - E_i = h\nu$$

where a and h are constants whose values are to be determined from other experimental or theoretical considerations. In fact, a may be obtained from the Rayleigh-Jeans law (2.22), which was thought to be a valid approximation under classical conditions, that is, at low frequencies:

$$a = 8\pi h/c^3$$

Substitution into (2.31) then gives the Planck law in terms of frequency,

$$u(\nu, T) = \frac{8\pi h\nu^3}{c^3} \left[\exp\left(\frac{h\nu}{kT} \right) - 1 \right]^{-1} \tag{2.32}$$

It has been stated that the Planck equation cannot be derived without an assumption which is not part of classical physics. It is not clear at which point in Einstein's derivation this assumption is introduced. It is not the existence of discrete energy levels, as the proof can be extended to the case of a continuum of levels. In fact, the non-classical assumption is that the atoms can exist in stable (stationary) states, with negligible time involved in the transition from one to another

2.6 THE MODERN THEORY OF THE RADIATION FIELD

The modern concepts used in the treatment of black-body radiation are in many ways very similar to those put forward by Planck. The Hamiltonian describing the radiation field is identical in form to that for an assembly of simple harmonic oscillators, echoing Planck's analysis of resonators in a cavity. Each mode k of angular frequency

$$\omega_k = 2\pi\nu_k$$

is formally equivalent to a single harmonic oscillator of energy

$$E = \left(n + \frac{1}{2}\right)\hbar\omega_k$$

The Hamiltonian for a simple harmonic oscillator is

$$\hat{H} = \frac{1}{2}(\hat{P}^2 + \omega_k^2 \hat{Q}^2)$$

where $\hat{P}$ and $\hat{Q}$ are the reduced forms of the momentum and displacement operators $\hat{p}$ and $\hat{q}$, with

$$\hat{P} = \frac{\hat{p}^2}{m} \qquad : \qquad \hat{Q} = m\hat{q}^2$$

where m is the mass of the particle and the momentum operator

$$\hat{p} = -ih\frac{\partial}{\partial\hat{q}}$$

For a mode of the radiation field characterised by the field vector $\boldsymbol{k}$ and the polarisation state $p_{\lambda k}$, the Hermitian operators $\hat{Q}_{k\lambda}$ and $\hat{P}_{k\lambda}$ obey commutation relationships

$$[\hat{Q}_{k\lambda}, \hat{P}_{k'\lambda'}] = ih\delta_{kk'}\delta_{\lambda\lambda'}$$
$$[\hat{Q}_{k\lambda}, \hat{Q}_{k'\lambda'}] = 0$$
$$[\hat{P}_{k\lambda}, \hat{P}_{k'\lambda'}] = 0$$

$\hat{Q}_{k\lambda}$ and $\hat{P}_{k\lambda}$ may be used to define the (non-Hermitian) annihilation and creation operators

$$\hat{a}_{k\lambda} = (2h\omega_k)^{-1/2}(\omega_k\hat{Q}_{k\lambda} + i\hat{P}_{k\lambda})$$
$$\hat{a}^*_{k\lambda} = (2h\omega_k)^{-1/2}(\omega_k\hat{Q}_{k\lambda} - i\hat{P}_{k\lambda})$$

which obey commutation relations

$$[\hat{a}_{k\lambda}, \hat{a}_{k'\lambda'}] = \delta_{kk'}\delta_{\lambda\lambda'}$$
$$[\hat{a}_{k\lambda}, \hat{a}^*_{k'\lambda'}] = 0$$
$$[\hat{a}^*_{k\lambda}, \hat{a}^*_{k'\lambda'}] = 0$$

The Hamiltonian for the radiation field then becomes

$$\hat{H} = \sum_{k\lambda} \hat{H}_{k\lambda}$$

$$= \sum_{k\lambda} \hbar\omega_k \left(\hat{a}_{k\lambda}^* \hat{a}_{k\lambda} + \frac{1}{2} \right)$$

$$= \sum_{k\lambda} \hbar\omega_k \left(\hat{N}_{k\lambda} + \frac{1}{2} \right)$$

where $\hat{N}_{k\lambda}$ is known as the number operator for the mode (k, λ). The eigenvalues $n_{k\lambda}$ of $\hat{N}_{k\lambda}$ are the positive integers 0, 1, 2 ..., so that the total energy E is

$$E = \sum_{k\lambda} E_{k\lambda}$$

$$= \sum_{k\lambda} \hbar\omega_k \left(n_{k\lambda} + \frac{1}{2} \right) \tag{2.33}$$

The eigenvalues $n_{k\lambda}$ are the photon occupation numbers (see §2.4.2), the number of photons in the mode (k, λ), and the radiation field may be fully described by enumerating them. The creation and annihilation operators are so named, of course, because they act on a given eigenstate to respectively increase or decrease the occupation number of that mode by unity.

2.7   ZERO-POINT FLUCTUATIONS OF THE RADIATION FIELD

It would be expected from the form of the Stefan-Boltzmann and Planck laws that the radiation field would completely disappear within a closed cavity at absolute zero, that is, at a temperature of 0 K. This is equivalent to saying that the occupation numbers $n_{k\lambda}$ would all be zero. However, it will be seen from (2.33) that each mode retains an energy $\hbar\omega/2$, which is known as the zero-point energy. As the number of modes even in a cavity of finite size is infinite, the residual energy in an evacuated cavity at absolute zero is infinite. This may be regarded as a peculiarity of the quantum mechanical description, perhaps associated with the fact that since all the occupation numbers are zero and hence exactly known the Heisenberg uncertainty principle dictates that the conjugate quantities, or the fluctuations associated with them, are infinite. The rather unsatisfactory solution usually adopted to the problem is to note that most physical measurements deal with changes in energy, rather than the absolute level, so that we may 'renormalise' the energy scale to remove the infinite total zero-point energy.

Even so, the fluctuations of the zero-point radiation field remain, and can produce significant physical effects. It would be expected that the zero-point field would, like black-body radiation, be isotropic and homogeneous. Unlike a black-body field at a positive temperature, however, it should appear the same to observers moving relatively to one another, as there is no associated parameter which can change as a result. Mathematically, this means that the spectral distribution must be invariant to a Lorentz transformation. The only form of spectral distribution which demonstrates this characteristic is one varying with the cube of the frequency. This conclusion agrees with that calculated from the density of modes derived by Jeans (2.15), and the zero-point energy of each mode.

While it might be expected that a radiation field with infinite energy might be rather easy to detect, it turns out that its effects are in reality rather subtle. It is possible to base the interpretation of several atomic effects upon its presence. For example, spontaneous emission may be regarded as a form of stimulated emission from an excited atom by the zero-point field. In most cases, however, other alternative explanations may be employed, which also correctly predict the effects observed but do not depend upon the presence of the field. The clearest demonstration of the effects of the zero-point field is afforded by the prediction by Casimir that two electrically conducting but uncharged parallel plates would experience an attractive force in a vacuum from the radiation field around them. Experimental confirmation was provided in 1958 by the Dutch physicist Sparnaay. At very low temperatures, there is a residual force F given by

$$F = \frac{hA}{2d^4}$$

where h is Planck's constant, A is the area of the plates and d their separation. The form of this equation is that predicted from the spectral distribution of the zero-point field postulated above.

2.8 DEVIATIONS FROM PLANCK'S LAW

There are of course many practical problems in the construction of black-body cavities which aim to provide a close approximation to the ideal and which are required for the most critical measurements. Here we just note that the assumptions made in the theoretical derivation of Planck's law are not always valid, and may give rise to detectable errors. In particular, it should be noted that it has been assumed that the wavelengths of black-body radiation are much smaller than any critical dimension of the cavity. When this is not the case, the mode distribution can no longer be taken as continuous, and the effects of a discrete distribution must be estimated. These may not only affect the Planck and Stefan-Boltzmann laws, but may significantly increase the mean square fluctuations of the radiation field when the mode distribution is irregular.

As a result there are two requirements upon the dimensions of the cavity for the effects to remain negligible for most radiometric applications. The first is that the smallest significant dimension of the cavity, d, should be about one hundred times greater than the wavelength being measured. While this condition is almost always found in conventional radiation thermometry, it is not the case of course at millimetre and microwave wavelengths. The second is that the spectral bandwidth of the detection system should obey the condition

$$\frac{\Delta\lambda}{\lambda} \gg \frac{\lambda}{d}$$

Again, it is unlikely in conventional radiation thermometry that this requirement will not be met, but it should be borne in mind if very narrow bandwidth interference filters are to be used for some reason.

The mode density $D(\nu)$ has been calculated by Baltes and colleagues [8][9][10] for frequencies where these conditions are not satisfied, and where the cavities are of relatively simple geometry and have smooth perfectly reflecting walls. Very large fluctuations in the mode density may be found for symmetrical cavities, for example spheres and cubes, where the mode structure is degenerate. In practical black-body cavities, of course, the roughness of the walls and the use of baffles inside the cavity would largely remove this degeneracy. Where the deviations are not too great, Baltes has shown by smoothing out the irregularities in the mode structure that the approximation

$$D(\nu) = D_0(\nu)\left[1 - \frac{\Gamma}{8\pi V}\left(\frac{c}{\nu}\right)^2\right]$$

may be useful. D_0 is the limiting mode density calculated for cavities where the conditions above are satisfied. V is the volume of the cavity, and Γ is a geometrical factor equal to three times the length of a side for a cube, six times the radius for a sphere, and so on.

The effect upon the Stefan-Boltzmann equation, for example, is such that the total energy within the cavity is more accurately given by

$$E(T) = \left(\frac{4\sigma}{c}\right)VT^4 - \frac{\pi^2\Gamma}{6hc}(kT)^2 + \frac{kT}{2} + \cdots$$

The conventional equation is adequate when the product

$$TV^{1/3} \gg hc/k \approx 1.5\,\mathrm{cm\,K}$$

2.9   COHERENCE PROPERTIES OF THERMAL RADIATION

The classical illustration of coherence in optical radiation often refers to simple effects such as the decrease in fringe contrast in an interferometer with path difference. This approach has its limitations; the coherence time τ_c of monochromatic radiation depends only upon the bandwidth, and cannot in principle distinguish that derived from chaotic sources from coherent laser radiation, although it is in practice much simpler to achieve a very narrow bandwidth with the latter.

This spectral aspect of coherence is described by the normalized first-order correlation function of the radiation field

$$g^{(1)}(\tau) = \frac{\langle A(r,t)A(r,t+\tau)\rangle}{\langle A(r,t)\rangle^2}$$

But the higher order coherence functions are not the same for coherent and chaotic sources. If we designate the power in the radiation field by $P(r, t)$, then the second-order field correlation function is

$$g^{(2)}(\tau) = \frac{\langle P(r, t) P(r, t + \tau) \rangle}{\langle P(r, t) \rangle^2}$$

For a laser which has no intensity fluctuations, $g^{(1)}(\tau)$ is equal to unity, and the photon statistics are Poissonian. For a chaotic source which, as described in §2.4.2, has intensity fluctuations from the variations in the photon number in each mode of the radiation field, it may be shown that

$$g^{(2)}(\tau) = 1 + |g^{(1)}(\tau)|^2$$

In the pure single mode case where the probability distribution is given by (2.26) for Bose-Einstein statistics, $g^{(2)}(\tau)$ is equal to 2. The result is manifest in photon bunching and the Hanbury-Brown and Twiss effect. For many modes thermal radiation may still show residual coherence effects over very short time intervals.

2.10 SPECTRAL RADIANCE

The theoretical derivations of the Planck law have given the energy density $u(\nu, T)$ in terms of frequency. In practice, we would much prefer to consider the flux emitted at a given wavelength and in a given direction, that is, the spectral radiance $L_\lambda(\lambda, T)$. This concept is important in radiation thermometry because it is invariant along a beam travelling in a passive lossless isotropic medium. Strictly speaking, if the beam enters a medium with a different refractive index n, the quantity L_λ/n^2 is conserved, but in the vast majority of practical cases the ray begins and finishes its journey in air. In general, of course, energy will be lost from the beam by absorption in the intervening medium, and by scattering and reflection at the surfaces of optical components. However, these can be allowed for with corrections which are usually independent of the geometry of the system.

Equation (2.24) for the energy density may readily be converted to the spectral radiance with (2.4), and to dependence on wavelength with (2.1). Up to this point, the effect of the refractive index of the medium through which the radiation travels has not been included in these equations; if the correction to the speed of light is made at the appropriate points in this chapter, the spectral radiance is given by

$$L_\lambda(\lambda, T) = \frac{2hc^2}{\pi^2 \lambda^5} \left[\exp\left(\frac{hc}{nk\lambda T} \right) - 1 \right]^{-1} \tag{2.34}$$

Experience has shown that there are a number of traps for the unwary in the evaluation of this equation in real situations. The first is that λ here is the wavelength of the radiation in the medium, while $n\lambda$ is the wavelength *in vacuo*. The transmission curves of colour and interference filters are normally obtained with monochromators whose wavelength scale is calibrated against convenient spectral lines. In the past, errors have been made because the air wavelengths of these lines are usually quoted in the reference books, and no correction for the refractive index of air has been applied. The errors are significant but not large; in one determination of the freezing point of platinum at 2040 K, for example, it was about 0.3 K [11].

The Planck equation is more commonly written in terms of the first and second radiation constants c_1 and c_2, whose values are usually taken from the list of recommended values produced by the CODATA Task Group on Fundamental Constants [12]:

$$c_1 = 2\pi hc^3 = (3.741\,771\,790 \pm 0.000\,000\,046) \times 10^{-16} \mathrm{W\,m}^2$$
$$c_2 = hc/k = 1.438\,777\,36 \times 10^{-2} \mathrm{m\,K}$$

It should be noted that c_1 contains a factor of π in addition to those expected from (2.34) above. This presumably arose because it is defined from the equation for the spectral exitance rather than the more commonly used spectral radiance. In addition, c_2 is normally taken as 1.4388×10^{-2}m K, because this value is specified in the definition of the International Temperature Scale. The difference does not lead to a temperature error that would be detectable in practice.

In terms of c_1 and c_2, the Planck equation for the spectral radiance becomes

$$L_\lambda(\lambda, T) = \frac{c_1}{\pi} \lambda^{-5} [\exp(c_2/\lambda T) - 1]^{-1} \tag{2.35}$$

In most applications, the ratio of two signals is calculated and only the second radiation constant is important. When regular use is made of the equation, it may be convenient to express c_2 in terms of the normal wavelength unit, that is, as $1.4388 \times 10^4 \mu\mathrm{m\,K}$ in the infrared and $1.4388 \times 10^7 \mathrm{nm\,K}$ in the visible. It is occasionally necessary to estimate the actual amount of power to be detected by a radiation thermometer and then the above equation must be evaluated in its entirety. The units of c_1 conceal the fact that the linear part arises from three separate quantities – the emitting area, the wavelength and the optical bandwidth. As the latter does not appear explicitly in (2.35), some confusion may occur.

In this case it is best to write down all of the relevant factors and convert all of those with length dimensions to metres first. To assist in checking the validity of such calculations, Figure 2.6 gives a single curve, scaled along both axes, which enables a rough answer to be quickly obtained. The wavelength-temperature product λT is measured off along the horizontal axis and a line drawn up to the curve. The corresponding reading on the vertical axis is then multiplied by T^5 and by the optical bandwidth expressed in metres to obtain the spectral radiance in watts per square metre of emitting surface per steradian. For example, for a radiation thermomet with an operating wavelength of 660 nm and a bandwidth of 30 nm, the λT product at 2000 K is $660 \times 10^{-9} \times 2000 = 0.001\,32$ m K. The corresponding intercept on the vertical axis is 1.65×10^{-7}, so that the spectral radiance detected would be $1.65 \times 10^{-7} \times 2000^5 \times 30 \times 10^{-9} = 158$ W m^{-2} sr^{-1}.

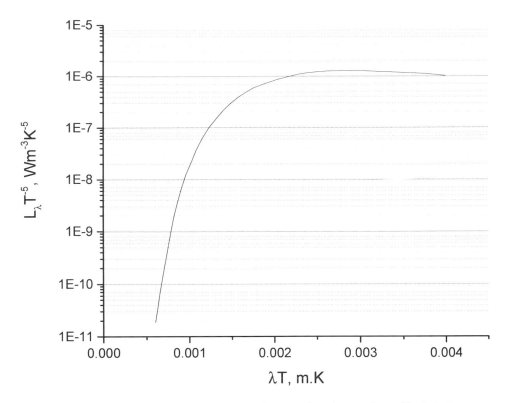

Figure 2.6: Single curve for the estimation of spectral radiance from black-body sources.

The wavelength corresponding to the maximum spectral radiance given by Wien's displacement law (2.9) may be obtained by differentiating (2.35) to find the maximum, and solving the resulting equation,

$$\frac{c_2}{c_3} = 5\left[1 - \exp\left(-\frac{c_2}{c_3}\right)\right]$$

which gives a numerical value for c_3 of 2897.7 μm K.

The approximations to the Planck distribution which are valid for limited ranges of the product λT may be derived without difficulty from the exact expression. For example, the Rayleigh-Jeans law may be obtained from (2.24) by expanding the exponential term as a series:

$$\exp\left(\frac{h\nu}{kT}\right) - 1 = \frac{h\nu}{kT} + \frac{1}{2!}\left(\frac{h\nu}{kT}\right)^2 + \cdots$$

If $(h\nu/kT)$ is small, only the first term is important. This law is of historical interest only, and has very little practical value in radiation thermometry. This is not true for the Wien approximation

$$L_\lambda(\lambda, T) = c_1 \lambda^{-5} \exp\left(-\frac{c_2}{\lambda T}\right)$$

obtained simply by neglecting the '−1' in the denominator of the Planck law. This is accurate to within 1% for values of the product λT below $3000\,\mu m\,K$, that is, from short wavelengths up to the maximum in the Planck curve at the temperature specified. As most radiation thermometry is carried out under these conditions, the mathematical simplicity of the Wien equation is of great value in the analysis of problems. For example, by taking logs and differentiating, we may show that the uncertainties in the spectral radiance, the temperature and the wavelength may be related by

$$\frac{\Delta L_\lambda}{L_\lambda} = \left(\frac{c_2}{\lambda T}\right)\frac{\Delta T}{T} + \left(\frac{c_2}{\lambda T} - 5\right)\frac{\Delta \lambda}{\lambda}$$

The combination $c_2/\lambda T$ occurs so frequently in radiation thermometry analysis that I have termed it 'the Planck parameter', denoted by the symbol 'p'. It was at one time common to approximate the variation of spectral radiance with temperature by a power law, that is,

$$L_\lambda(\lambda, T) \propto T^n$$

It is relatively simple to show that for small changes the exponent is in fact the Planck parameter p. As p is itself a strong function of temperature, the power law approximation is only valid over a very limited range, and its use is discouraged in favour of the Wien equation. Finally, we note that if the spectral exitance of a blackbody is integrated over all wavelengths, the total exitance is obtained as a function of temperature, that is, the Stefan-Boltzmann law. Making the substitution

$$x = \frac{h\nu}{kT}$$

and using the mathematical relationship

$$\int_0^\infty \frac{x^3}{e^x - 1}dx = \frac{4}{15}$$

we obtain

$$M(T) = \left[\frac{2\pi^5 k^4}{15c^2 h^3}\right]T^4$$

enabling the Stefan-Boltzmann constant to be evaluated in terms of other fundamental constants[2]. Its value, to an accuracy adequate for most purposes, is $5.67 \times 10^{-8}\,W\,m^{-2}\,K^{-4}$, which must make it the easiest fundamental constant to remember!

---

[2]The value of c is fixed by the S.I. definition of the metre and the definition of the second. The stated intention of the Bureau International des Poids et Mesures (BIPM) is to specify the kelvin by fixing the value of the Boltzmann constant with zero uncertainty. Also, the kilogram is to be specified by similarly fixing the Planck constant. The Stefan-Boltzmann constant will therefore also be fixed with zero uncertainty.

CHAPTER 3

Characteristics of surfaces

3.1 GENERAL CHARACTERISTICS OF SURFACES

It was noted in the previous chapter that the interaction of thermal radiation with solids and liquids involved a wide range of physical effects. However, in the discussion of the fundamental laws of radiation thermometry in that chapter only two ideal surfaces – a perfect reflector and a black-body – were necessary. Real surfaces are of course more complex than these, and in this chapter we shall describe the optical properties required to characterise them.

The importance of these properties may be indicated by considering the measurement system shown in Figure 3.1. The radiation thermometer will typically detect the thermal radiation from the target area on the surface A within a restricted range of wavelengths. In general, the relationship between the detected signal and the temperature will have been determined by calibrating the radiation thermometer against a reference source of thermal radiation, for example a black-body cavity whose temperature may be measured with a thermocouple. The radiance emitted by the real surface will be less than that from a black-body at the same temperature by a factor which is known as the emittance or emissivity[1] ϵ of the surface, and the reading of the radiation thermometer will therefore be lower than the true temperature of the surface. As ϵ may lie anywhere in the range from about 0.01 up to 1, the error may be severe and a reasonable estimate of the emittance is required if the temperature uncertainty is to be acceptable. Commercial radiation thermometers enable estimated values for the surface under test to be entered so that an automatic correction to the reading may be made.

It is also possible for radiation from other sources in the vicinity of the target surface to be reflected or scattered into the thermometer, increasing the observed

---

[1]These terms are often considered to be interchangeable. It has been suggested in the USA that the termination 'ance' to the name of an optical quantity should indicate that it applies to a real, necessarily imperfect, surface, while that of 'ivity' implies reference to an idealised smooth uncontaminated surface. The emissivity, for example, would be determined by the material of which the surface was composed alone, while the emittance would be affected by the method of preparation of the surface, its roughness and the presence of thin films of oxides and other contaminating compounds.

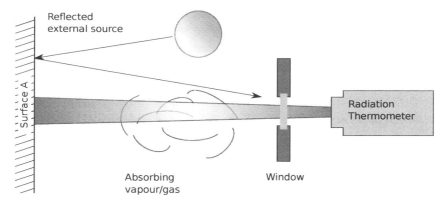

Figure 3.1: Radiation thermometry measurement.

temperature. The magnitude of the error produced depends not only on the size and temperature of the external source and the reflectance of the target area, but also upon its directional characteristics, that is, upon the proportions specularly or diffusely reflected towards the thermometer. If the surface is a black-body, all incident radiation is absorbed and no error can result.

In passing, we may note that some of the thermal radiation may be absorbed or scattered out of the optical path between the target area and the radiation thermometer, again reducing the apparent temperature. These effects may be produced by gases, vapours or smoke in the sight path, or by windows through which the surface is observed and which form part of a shield designed to protect either the surface or the thermometer. They will be described in more detail in later sections.

From this example, it is clearly advisable to determine as far as possible the optical properties of the surface that may affect the reading of the thermometer. This chapter sets out to indicate which properties are relevant, and to indicate their general trends with temperature, wavelength, viewing direction and so on.

3.1.1   Definition of optical properties

In this section we will define the parameters that describe the main optical properties of materials, that is, the proportions of the incident radiation that are reflected, absorbed or transmitted. The relationship between these parameters and the amount of thermal radiation emitted by an object will then be developed. We first consider a monochromatic beam of flux Φ_i illuminating an area from a direction (θ_i, ϕ_i). The area ΔA is assumed to be sufficiently small that it may be considered flat and as having uniform characteristics. The basic geometry is shown in Figure 3.2. θ_i is the zenith angle from the normal ON to ΔA, while the azimuthal angle ϕ_i is measured with respect to some arbitrary axis OA in the plane of ΔA. Some of the reflected radiation may be observed within a solid angle $\Delta \omega_r$ in the direction (θ_r, ϕ_r) above the upper surface of the block; if the material of the block absorbs weakly at the wavelength λ, that is, it is transparent or translucent, transmitted radiation may be detected emerging from the lower surface in a direction (θ_t, ϕ_t).

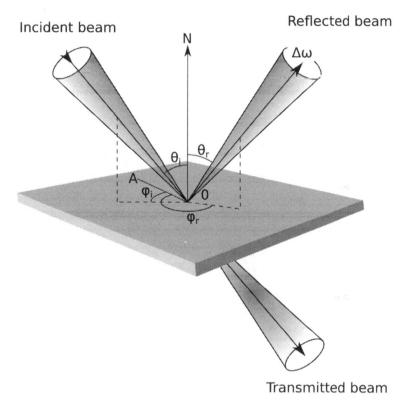

Figure 3.2: Geometry for optical properties.

In the absence of fluorescence or non-linear optical effects, the wavelength of the reflected or transmitted radiation remains unchanged. The radiation absorbed is however converted into thermal energy transferred to the block. While the fraction absorbed could be measured by calorimetric methods, it is usually obtained from the other optical quantities, for example, from (3.7).

We shall follow the convention of indicating the dependence of the optical properties upon other factors by placing these within parentheses after the symbol for the quantity and separated by semicolons. Related variables within a factor, such as the zenith and azimuth angles giving the direction of a beam, are separated by commas. For example, the absorbed flux Φ_a from the incident beam depends upon the wavelength and the direction of the incident beam. The absorptance for a material at temperature T may therefore be defined by the equation

$$\alpha(\lambda; \theta_i, \phi_i; T) = \frac{\Phi_\alpha(\lambda; \theta_i, \phi_i; T)}{\Phi_i} \tag{3.1}$$

where the subscript i indicates a property of the incident beam. As it may be taken that all optical properties are dependent to a greater or lesser extent upon T, the notation will be simplified in the following sections by omitting this term from within the parentheses.

The situation is more complicated in the case of reflected radiation, as we may wish to determine the proportion reflected in a particular direction. The reflected flux $\Phi_r(\lambda; \theta_i, \phi_i; \theta_r, \phi_r; \Delta\omega_r)$ depends not only upon the directions of incidence and reflection, but also upon the solid angle within which the reflected flux is collected. If we define a reflectance simply in terms of the ratio of the reflected to the incident flux, Φ_r/Φ_i, it will be dimensionless but its value will depend on this solid angle. This is unsatisfactory, and it is preferable to normalise the reflected flux to give the proportion per unit solid angle. The bidirectional reflectance is therefore defined by the equation

$$\rho'(\lambda; \theta_i, \phi_i; \theta_r, \phi_r) = \frac{\Phi_r(\lambda; \theta_i, \phi_i; \theta_r, \phi_r)}{\Phi_i \cos\theta_r \Delta\Omega_r} \tag{3.2}$$

and has the units sr^{-1}.

An important theorem in the measurement of the properties of real surfaces is the Helmholtz reciprocity theorem, which states that

$$\rho'(\lambda; \theta_i, \phi_i; \theta_r, \phi_r) = \rho'(\lambda; \theta_r, \phi_r; \theta_i, \phi_i) \tag{3.3}$$

that is, that the bidirectional reflectance is unchanged if the directions of the incident and reflected beams (θ_i, ϕ_i) and (θ_r, ϕ_r) are interchanged. This theorem is valid for most surfaces in radiation thermometry, and only fails for unusual situations, for example, where fluorescence takes place, or the surface is an active polariser.

Although ideally we would like to determine the bidirectional reflectance for all possible directions, it is a complex function and would require a great deal of experimental data. In addition, it will depend not only upon the material concerned, but also upon the state of the surface, especially its roughness. In practice we often have to be satisfied with a directional-hemispherical reflectance which gives the proportion of the incident beam reflected or scattered in all directions, that is,

$$\rho(\lambda; \theta_i, \phi_i; 2\pi) = \frac{\Phi_r(\lambda; \theta_i, \phi_i; 2\pi)}{\Phi_i} \tag{3.4}$$

where the '2π' indicates that the reflected radiation is detected over the complete hemisphere. This quantity is related to the bidirectional reflectance by

$$\rho(\lambda; \theta_i, \phi_i; 2\pi) = \int_H \rho'(\lambda; \theta_i, \phi_i; \theta_r, \phi_r) \cos\theta_r d\omega_r \tag{3.5}$$

where the 'H' denotes integration over the hemisphere above the surface.

The quantitative description of the transmitted radiation is still more complicated in the general case. The first problem is that absorption of radiation usually follows an exponential law, so that the fraction transmitted depends upon the optical path length within the block, and hence upon its dimensions and the angle of incidence. More seriously, refraction, scattering and internal reflection within the solid have the result that the transmitted radiation does not emerge from a specific and well-defined point on the lower surface. Although the transmittance is often treated in the same way as the reflectance, as a bidirectional quantity, this is an adequate description in only a limited number of cases. It is fortunate, therefore, that most materials we shall deal with are opaque, that is, the absorption coefficient is so large that the proportion of transmitted radiation is negligible.

The most practical solution for materials which are not opaque is to define a directional-hemispherical transmittance along the lines of (3.4),

$$\tau(\lambda; \theta_i, \phi_i; 2\pi) = \frac{\Phi_t(\lambda; \theta_i, \phi_i; 2\pi)}{\Phi_i} \tag{3.6}$$

where the flux Φ_t is integrated over all directions and over the whole surface through which the transmitted flux emerges. As the incident flux can only be absorbed, reflected or transmitted, the law of conservation of energy requires that

$$\Phi_a(\lambda; \theta_i, \phi_i) + \Phi_r(\lambda; \theta_i, \phi_i; 2\pi) + \Phi_t(\lambda; \theta_i, \phi_i; 2\pi) = \Phi_i$$

and hence that

$$\alpha(\lambda; \theta_i, \phi_i) + \rho(\lambda; \theta_i, \phi_i; 2\pi) + \tau(\lambda; \theta_i, \phi_i; 2\pi) = 1 \tag{3.7}$$

As noted above, the absorption coefficients of visible and infrared radiation inside many solids can be extremely high, and the effective transmittance is zero. Thus for opaque solids, this equation reduces to

$$\rho(\lambda; \theta_i, \phi_i; 2\pi) + \alpha(\lambda; \theta_i, \phi_i) = 1 \tag{3.8}$$

3.1.2 Kirchhoff's law

The interaction of a beam of radiation with a surface is independent of the simultaneous process of thermal emission, unless the energy absorbed is sufficient to raise its temperature significantly. The amount of thermal energy radiated in a specified direction is characterised by the emittance, which may be defined as the ratio of the radiance from the surface to that from a black-body at the same temperature, that is,

$$\epsilon(\lambda; \theta_i, \phi_i; T) = \frac{L_\lambda(\lambda; \theta_i, \phi_i; T)}{L\lambda^B(\lambda; T)} \tag{3.9}$$

The superscipt 'B' indicates black-body conditions; it was demonstrated in §2.2.2 that the radiance from a black-body surface is independent of direction, and this characteristic of course applies equally to the spectral radiance.

The emittance can be related to the other optical properties of the surface with Kirchhoff's law. The most general form of this law may be derived by considering as before a flat disc within a black-body cavity, but with the black-body surface replaced by an opaque material of arbitrary emittance. Applying the geometrical notation shown in Figure 3.2 (with the transmitted component removed), the principle of detailed balance notes that the flux leaving the disc surface in the direction must be balanced by that reaching it from the radiation field within the cavity. From the definition of radiance given in §2.2.1, the flux reaching the surface may be written as

$$d\Phi_r = L_\lambda^B(\lambda; T) \cos\theta_r dA\delta\omega_r \tag{3.10}$$

The flux leaving the surface is the sum of that thermally emitted and that reflected in the direction (θ_r, ϕ_r) from the uniform illumination of the disc by the radiation field within the cavity. The component reaching the disc along some arbitrary direction (θ_i, ϕ_i) is given by

$$d\Phi_i = L_\lambda^B(\lambda; T) \cos\theta_i dA\delta\omega_r$$

The part of this reflected from the disc into the solid angle $d\omega_r$ in the direction (θ_r, ϕ_r) is obtained from the definition of the bidirectional reflectance in (3.2),

$$d\Phi_i \rho'(\lambda; \theta_i, \phi_i; \theta_r, \phi_r) \cos\theta_r \, d\omega_r$$

After rearranging the terms within this expression, the total reflected in this direction from the hemispherical illumination of the disc may be written

$$d\Phi_r = L_\lambda^B(\lambda; T) \cos\theta_r \, dA \, d\omega_r \int_H \rho'(\lambda; \theta_i, \phi_i; \theta_r, \phi_r) \cos\omega_i d\omega_i$$

Applying the Helmholtz reciprocity theorem yields an integral of the form given in (3.5), and hence

$$d\Phi_r = L_\lambda^B(\lambda; T) \cos\theta_r \, dA \, d\omega_r \rho(\lambda; \theta_r, \phi_r; 2\pi)$$

As the flux from thermal emission in the direction (θ_r, ϕ_r) may be written from (3.9) as

$$d\Phi_e = \epsilon(\lambda; \theta_r, \phi_r) L^B_\lambda(\lambda; T) \cos\theta_r \, dA \, d\omega_r$$

equating the sum of the exitent fluxes with the incident flux from (3.10) and dropping the redundant subscripts gives

$$\epsilon(\lambda; \theta, \phi) + \rho(\lambda; \theta, \phi; 2\pi) = 1 \tag{3.11}$$

and hence from (3.8)

$$\epsilon(\lambda; \theta, \phi) = \alpha(\lambda; \theta, \phi) \tag{3.12}$$

which is Kirchhoff's law.

Equation (3.11) is of considerable practical importance as it enables the emittance to be derived directly from measurements of the directional reflectance. At temperatures around ambient, it is often a much simpler matter to measure reflectances over the necessary range of angles and wavelengths than the corresponding values of emittance. The main reasons for this are that high intensities may be employed in the reflectance measurements, giving good signal-to-noise ratios, and that the incident beam may also be chopped so that the reflected radiation can be distinguished from thermally emitted or stray radiation. It may be noted in passing that (3.7) implies that the maximum possible value for the absorptance, reflectance and transmittance is unity. Equation (3.12) also reflects the fact that a black-body surface, which absorbs all incident radiation, has an absorptance and hence an emittance of unity. It follows that all real surfaces emit less radiation than a black-body at the same temperature. This characteristic is sometimes used to define a black-body surface.

Kirchhoff's law is always derived for a surface in thermal equilibrium with the radiation field in a cavity at the same temperature. In most applications, the surface will be at a much higher temperature than its surroundings, and it is therefore not in equilibrium with the ambient radiation field. Under these circumstances, the universal validity of the law may be questioned. Discussion of this point has been complicated by a misinterpretation of the significance of Einstein's derivation of the Planck radiation law, outlined in Chapter 2. If the radiation emitted by a surface is the sum of that spontaneously emitted and the stimulated emission induced by the presence of the surrounding radiation field, then the emittance would indeed depend upon the energy density of the field. The effect would be expected to be quite significant when the product λT is large, that is, in the long wavelength tail of the Planck spectral distribution. The question has been discussed at length by Baltes [13], who pointed out that it is incorrect to separate the processes of stimulated absorption and emission. Stimulated emission should be regarded as a 'negative' component of absorption, while the emittance in Kirchhoff's law corresponds to the spontaneous radiative process alone, which is independent of the local radiation field.

To this point we have retained the functional dependence of the optical properties upon the main variables, particularly the wavelength and the directions of the incident or emitted radiation. In radiometric calculations these forms are the most important; the majority of radiation thermometers accept radiation over a rather limited wavelength range in the Planck distribution, have a narrow field of view in angular terms, and usually are directed normally to the target surface. In other applications, especially for heat transfer calculations, it is convenient to use optical quantities which involve the total radiative flux, summed over all wavelengths or over all directions.

There are no great difficulties in defining and applying these variants of the emittance. For example, the total emittance which includes radiation at all wavelengths may be defined in terms of the ratio of the actual radiance (not the spectral radiance) in the required direction to that from a black-body source,

$$\epsilon(\theta, \phi; T) = \frac{L(\theta, \phi; T)}{L^B(T)} \tag{3.13}$$

and may be calculated by integrating the weighted spectral emissivity in that direction over all wavelengths,

$$\epsilon(\theta, \phi; T) = \frac{\int_0^\infty \epsilon(\lambda; \theta, \phi; T) L_\lambda(\lambda; \theta\phi; T)\, d\lambda}{L^B(T)}$$

Similarly, hemispherical spectral and total emittances, involving the radiation emitted over all angles, may be defined using the spectral and total exitances

$$\epsilon(\lambda; 2\pi; T) = \frac{M_\lambda(\lambda; T)}{M_\lambda^B(\lambda; T)}$$

$$\epsilon(2\pi; T) = \frac{M(T)}{M^B(T)} \tag{3.14}$$

and these may also be expressed as weighted integrals of the spectral emittance.

However, the definition and measurement of the corresponding forms of the other optical quantities must be treated with some caution, as they involve radiation which is incident on the surface. Hemispherical and total quantities will depend respectively upon the angular and spectral distribution of this. We need to specify these distributions to ensure that the quantities so defined are single-valued; in addition, we would like to select distributions such that the relationships between the optical properties, and in particular Kirchhoff's law, remain valid. The condition for Kirchhoff's law to remain valid for hemispherical quantities, that is, those obtained when the incident rays illuminate the disc from all directions, is that the irradiation should be uniform. Under these conditions, we may write Kirchhoff's law in the form

$$\epsilon(\lambda; 2\pi; T) = \alpha(\lambda; 2\pi; T) \tag{3.15}$$

It may also be shown that Kirchhoff's law will be valid if the surface is perfectly diffuse, even if the spectral irradiance varies with direction. While this condition is approximated by rough surfaces, its assumption is best avoided unless supported by experimental data.

Similarly, it may be shown that Kirchhoff's law for total quantities, that is, those integrated over all wavelengths,

$$\epsilon(\theta, \phi; T) = \alpha(\theta, \phi; T)$$
$$\epsilon(2\pi; T) = \alpha(2\pi; T) \tag{3.16}$$

remains valid if the irradiance is proportional to the radiance from a black-body at the surface temperature T. An alternative requirement is that the surface is 'grey', that is, that the spectral emittance is independent of wavelength. While the first is difficult to arrange in many experimental systems, the second is satisfied by very few materials. It follows that only the total emittance is of any value for precise measurements; other total quantities should be treated with caution or avoided completely wherever possible.

The notation employed in this chapter is specific but rather cumbersome. Optical properties are rarely dependent upon the azimuth in particular, and this is often omitted. In later chapters it will prove advantageous to use abbreviated forms for those quantities commonly required in radiation thermometry, and these are listed in Table 3.1.

Table 3.1: Compressed notation for optical quantities

Quantity	Notation in full	Concise
Spectral emissivity normal to the surface	$\epsilon(\lambda; 0, 0; T)$	ϵ_λ
Spectral emittance at an angle θ to the normal	$\epsilon(\lambda; \theta, 0; T)$	ϵ_θ
Total emittance normal to the surface	$\epsilon(0, 0; T)$	ϵ_N
Total hemispherical emittance	$\epsilon(2\pi; T)$	ϵ_T
Spectral reflectivity normal to the surface	$\rho(\lambda; 0, 0; T)$	ρ_λ
Spectral reflectance at an angle θ to the normal	$\rho(\lambda; \theta, 0; T)$	ρ_θ

3.2 CALCULATED EMISSIVITY VALUES

We have remarked on the need for estimates of the optical characteristics of a surface in order to obtain reasonably accurate measurements of its temperature. Published data are generally unreliable without further investigation as the values can vary widely depending on sample preparation and condition. For the same reason measurements of a material's properties need to be specific to the material as it will be measured. In many of not most cases this will be impractical and so, in this section, we shall review the extent to which this information may be obtained from theories and models which relate the reflectivity and the emissivity to other, possibly more readily available, properties of the material of the surface. Initially we shall deal only with the ideal situation in which the material is homogeneous and the surface perfectly smooth. In this case, the reflection at the surface is completely specular.

It may readily be shown from Maxwell's equations that in a non-absorbing material, that is, a perfect dielectric, the refractive index n which is the ratio of the velocity of the wave *in vacuo* to that in the material, is given by

$$n = \frac{c_0}{c} = \left(\frac{\mu}{\mu_0} \frac{\gamma}{\gamma_0} \right)^{\frac{1}{2}}$$

where the subscript zero indicates the value of the quantity *in vacuo*. In general the magnetic permeability ratio μ/μ_0 is close to unity; the electrical permittivity ratio γ/γ_0 is known as the dielectric constant ζ or, more appropriately, as the dielectric function, since it may vary with frequency, temperature and other parameters.

If we consider an electromagnetic wave travelling in a material of refractive index n_1 and incident at an angle θ_i on the interface with a second material of index n_2 (Figure 3.3), the application of the boundary conditions necessary for the continuity of the electric and magnetic field components leads to simple relations for the reflected and refracted waves,

$$\theta_i = \theta_r = \theta$$
$$\frac{\sin \chi}{\sin \theta} = \frac{n_1}{n_2} \tag{3.17}$$

The latter is known generally as Snell's law. If the refractive index n_2 is less than n_1, at angles of incidence above a critical value there is no valid solution for χ, and the wave is completely reflected at the interface. This phenomenon is known as total internal reflection, and forms the basis of the containment of light within optical fibres.

For the two polarisation states, with the plane of polarisation containing the electric field vector parallel to and normal to the plane of the surface (indicated respectively by the superscripts 'p' and 'n'), the reflectivities are obtained from the square of the reflected component of the electric field intensity

$$\rho_\theta^p = \left[\frac{\cos \theta / \cos \chi - n_1/n_2}{\cos \theta / \cos \chi + n_1/n_2} \right]^2$$

$$= \frac{\tan^2(\theta - \chi)}{\tan^2(\theta + \chi)} \tag{3.18}$$

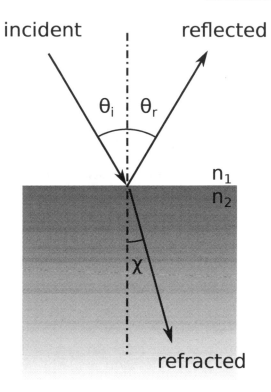

Figure 3.3: Electromagnetic wave at a surface.

and

$$\rho_\theta^n = \left[\frac{\cos\chi/\cos\theta - n_1/n_2}{\cos\chi/\cos\theta + n_1/n_2} \right]^2$$

$$= \frac{\sin^2(\theta - \chi)}{\sin^2(\theta + \chi)} \qquad (3.19)$$

These are known as the Fresnel equations. For unpolarised radiation, the reflectivity is given by the mean of the polarised components

$$\rho_\theta = \frac{\rho_\theta^p + \rho_\theta^n}{2} \qquad (3.20)$$

In radiation thermometry, the incident radiation is usually travelling in air or another medium of low refractive index, so that n_1 is effectively equal to unity. The wavelength dependence of the reflectivities is determined by that of the refractive index of the denser medium n_2.

Two conclusions may be derived directly from these equations. First, at a particular angle of incidence, the Brewster angle

$$\theta_B = \tan^{-1}(n_1/n_2) \qquad (3.21)$$

the reflectivity for the parallel component is zero, and the reflected beam contains only the normally polarised wave. While this effect may be employed as a polariser, more effective arrangements are in general use. If the incident beam is linearly polarised, however, it provides a method of passing radiation through a window with very little loss from reflection. Secondly, at normal incidence where $\theta = 0$

$$\rho_0 = \rho_0^p = \rho_0^n = \left(\frac{n_2 - n_1}{n_2 + n_1} \right)^2 \tag{3.22}$$

For example, the reflectivity in air of a single surface of glass which has a refractive index of 1.47 is around 0.04 (4%), whereas that of germanium which is used for optical components in the infrared, with n_2 around 4.0, is 36%.

The emissivities of dielectric surfaces may be obtained from the calculated reflectivities using Kirchhoff's law, but it should be noted that the results are only applicable when the sample is opaque or optically thick, that is, the fraction of any incident beam transmitted through the sample is negligible. In this case, the emissivity into air normal to the surface is

$$\epsilon_\lambda = \frac{4n_2}{(n_2 + 1)^2} \tag{3.23}$$

As the refractive index of most dielectrics is less than 5, it will be seen that the emissivity of these materials is generally quite high, greater than 0.5.

The Fresnel equations above have been used to calculate the variation with angle of the polarised components of the emissivity, and the combined emissivity, for a refractive index ratio n_2/n_1 equal to 2.5 and the results are plotted in Figure 3.4. It will be seen that the normally polarised component falls smoothly to zero, while the parallel component rises to a maximum value of unity at an angle of 70° before falling away, and in fact the unpolarised emissivity remains nearly constant out to this angle. Curves for the unpolarised emissivity for different ratios of n_2/n_1 are shown in Figure 3.5. The average values over a range of angles, including the hemispherical emissivity, may be obtained from these curves by numerical integration. They are of value in estimating the possible uncertainties arising from the use of radiation thermometers with large acceptance angles, particularly if they view the surface at an angle away from the normal.

3.2.1 Equations for strongly absorbing materials

If the electromagnetic wave is attenuated exponentially as it passes through a given material, the solutions of Maxwell's equations indicate that the refractive index may be treated as complex, that is, the real value n for a perfect dielectric should be replaced by $\tilde{n}$, where

$$\tilde{n} = n - i\kappa$$

and κ is known as the extinction coefficient. For metals attenuation is very large, typically around $10 \times 10^7 \, \mathrm{m}^{-1}$ for visible wavelengths, which indicates that the penetration depth of the incident radiation is only of the order of a wavelength.

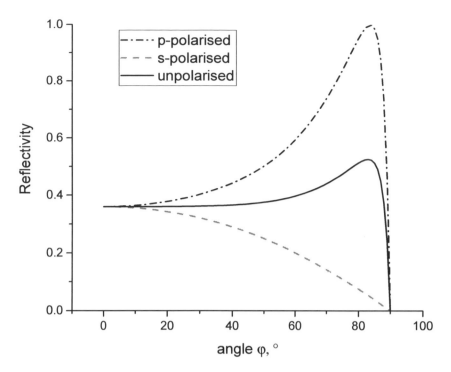

Figure 3.4: Polarised components of emissivity from Fresnel equations for dielectrics with
$n_1 = 1$ *and* $n_2 = 2.5$.

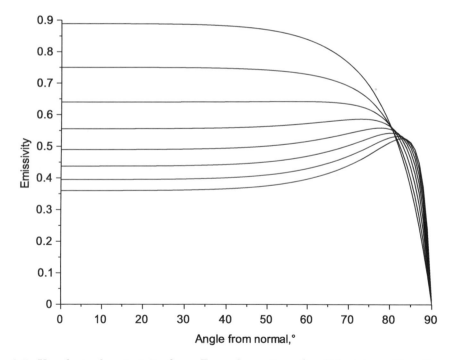

Figure 3.5: Unpolarised emissivity from Fresnel equations for dielectrics with ratio $n_2/n_1 =$
2 *to* $n_2/n_1 = 9$. *Increasing the ratio causes a reduction of emissivity for angles from normal*
to 40°.

Fortunately, it is possible to extend the equations derived for a perfect dielectric to cover absorbing materials with the simple procedure of replacing n wherever it occurs by $\tilde{n}$; where a function containing n is squared, the corresponding complex function of $\tilde{n}$ is multiplied by its complex conjugate in the usual way. For example, if we apply the procedure to (3.22) for the reflectivity at normal incidence, but now consider that the second material is absorbing with an extinction coefficient κ_2, the equation becomes

$$\rho_N = \frac{(n_2 - i\kappa_2) - n_1}{(n_2 - i\kappa_2) + n_1} \cdot \frac{(n_2 + i\kappa_2) - n_1}{(n_2 + i\kappa_2) + n_1}$$

$$= \frac{(n_2 + n_1)^2 + \kappa_2^2}{(n_2 - n_1)^2 + \kappa_2^2} \tag{3.24}$$

In the more general case where the beam is incident on the interface at an angle θ, the procedure remains simple in principle, but the algebra becomes tedious in practice. Again assuming that only the second material is absorbing, the reflectivities for the two polarisation states are given by

$$\rho_\theta^n = \frac{a^2 + b^2 + \cos^2\theta - 2a\cos\theta}{a^2 + b^2 + \cos^2\theta + 2a\cos\theta} \tag{3.25}$$

$$\rho_\theta^p = \frac{a^2 + b^2 + \sin^2\theta\tan^2\theta - 2a\sin\theta\tan\theta}{a^2 + b^2 + \sin^2\theta\tan^2\theta - 2a\sin\theta\tan\theta} \tag{3.26}$$

where

$$c = n_2^2 - \kappa_2^2 - n_1^2\sin^2\theta$$
$$2a^2 = (c^2 + 4n_2^2\kappa_2^2)^{\frac{1}{2}} + c$$
$$2b^2 = (c^2 + 4n_2^2\kappa_2^2)^{\frac{1}{2}} - c$$

For metals, where n_2 and κ_2 are often large, the angular term in the expression for c may be neglected. This implies that χ is small, and leads to much simpler expressions for the reflectivities:

$$\rho_\theta^p = \frac{(n_2\cos\theta - n_1)^2 + \kappa_2^2\cos^2\theta}{(n_2\cos\theta + n_1)^2 + \kappa_2^2\cos^2\theta}$$

$$\rho_\theta^n = \frac{(n_2 - n_1\cos\theta)^2 + \kappa_2^2}{(n_2 + n_1\cos\theta)^2 + \kappa_2^2} \tag{3.27}$$

Figure 3.6 shows the emissivity derived from these equations for a tungsten surface at a wavelength of 655 nm plotted as a function of angle for both polarisations, together with the combined emissivity. At small angles, up to about 10°, the emitted radiation is essentially unpolarised, but above this the degree of polarisation increases rapidly as a result of the very different shapes of the curves for the two

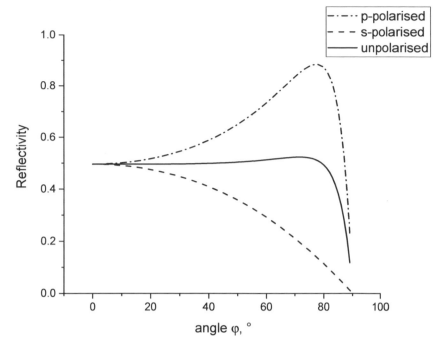

Figure 3.6: Emissivity for tungsten from Fresnel equations.

polarisation states. In addition, the total emitted radiance does not obey Lambert's law, but shows a maximum at 75° before dropping to zero.

The normal spectral emissivity ϵ_λ into air is more readily calculated from (3.23)

$$\epsilon_\lambda = \frac{4n}{(n+1)^2 + \kappa^2} \tag{3.28}$$

In the case of tungsten the value obtained is quite close to that measured directly. In general there are unfortunately two major problems with the estimation of emissivities from the optical constants n and κ using the equations above. The first is that the accuracy of the results is unpredictable. A comparison of calculated and experimental values is presented in Table 3.2 for metallic elements. In some cases the agreement is good, but in others differences of a factor of two or more are found, for no obvious reason. The second problem is that optical constants for only a few materials are well established. In most cases there are significant differences between the values found by different groups, and with different techniques. The main advantage of the Fresnel equations is that they provide a reasonably reliable indication of the variation of the optical properties with angle, and this information is of value in a number of radiometric applications.

3.2.2 The Drude free-electron theory

A simple theory which apparently proved successful in inter-relating the electrical, thermal and optical properties of metals was published as long ago as 1898 by

Table 3.2: Comparing emissivity calculated using (3.28) taking values for optical constants at 1 μm with typical measured values

Metal	Emissivity from (3.28)	Typical measured emissivity
Gold	0.03	0.3
Titanium	0.39	0.5-0.75
Aluminium	0.07	0.1-0.2
Tungsten	0.44	0.3-0.4
Iron	0.37	0.35
Magnesium	0.06	0.3-0.8
Molybdenum	0.36	0.25-0.35
Tin	0.19	0.25
Zinc	0.16	0.5

Drude. The conduction electrons are assumed to form a 'gas' which is to a large extent free of the positive ions in the lattice. The differential equation describing the interaction between the electron gas and an electromagnetic wave of angular frequency ω includes a damping term from the interaction of the electrons and the lattice which is characterised by a relaxation time, τ_R. The theory gives expressions for the real and imaginary parts of the dielectric function ζ:

$$\zeta_1 = \zeta_0 - \frac{\omega_P^2 \tau_R^2}{1 + \omega^2 \tau_R^2}$$
$$\zeta_2 = \frac{\zeta_0 - \zeta_1}{\omega \tau_R} \tag{3.29}$$

In principle, ζ_0 should be unity in value, but it may need to be increased by a term representing the contributions from the interband transitions which usually occur at high frequencies, that is, at shorter wavelengths than the greater part of the thermal distribution. ω_P is known as the plasma frequency, the frequency at which plasma oscillations may take place in the electron gas when driven by the electromagnetic wave. Its value is given by

$$\omega_P^2 = \frac{e^2 N_e}{m_e \epsilon_0} \tag{3.30}$$

where N_e is the number density of the conduction electrons, and m_e is their effective or 'optical' mass. The plasma frequency may alternatively be written in terms of the ratio of the radius a_s of the free electron sphere, defined by

$$N_e \left(\frac{4\pi a_s^3}{3} \right) = 1$$

to the Bohr radius, a_0. Values of this ratio are available for many metals, and are typically around 3-4, giving a plasma frequency of about 10^{16} radians per second.

Once the dielectric function is known, the optical constants may be calculated from

$$\zeta = \tilde{n}^2 \qquad \text{or} \qquad \zeta_1 + i\zeta_2 = (n - i\kappa)^2 \tag{3.31}$$

and hence the emissivity from (3.28). A number of simple approximations may be derived from the Drude equation for the spectral and total emissivities at long wavelengths, where the interband transitions may be neglected. When $\omega\tau_R \ll 1$, the solution for the optical constants reduces to

$$n^2 = \kappa^2 = \frac{\omega_P^2 \tau_R}{2\omega} \tag{3.32}$$

These may be related to the d.c. electrical resistivity of the material through

$$\omega_P^2 \tau_R = \frac{\mu_0 c_0^2}{r_e} \tag{3.33}$$

where r_e has units of $\Omega\,\text{m}$. If we substitute the condition $n = \kappa$ into (3.28) for the normal spectral emissivity and expand the result as a series, we obtain the Hagen-Rubens equation

$$\begin{aligned}
\epsilon_{\lambda,N} &\approx \frac{2}{n} - \frac{2}{n^2} + \cdots \\
&= 0.365(r_e/\lambda)^{\frac{1}{2}} - 0.0667(r_e/\lambda) + \cdots
\end{aligned} \tag{3.34}$$

where the wavelength is given in metres. Although this expression would only be expected to hold at longer wavelengths, above 5 μm, it would clearly still be of considerable value if valid, as it relates a quantity, the spectral emissivity, which is difficult to measure accurately, with the readily and precisely measurable d.c. resistivity of a metal.

Figure 3.7 shows the spectral emissivity of tungsten plotted against wavelength for different temperatures. The curves were calculated from the measured resistivity of tungsten at each temperature using (3.34), while the data points are taken from optical measurements of the emissivity giben in [14]. It will be seen that agreement is moderate above a wavelength of 3 μm, although in percentage terms the differences are not negligible. At short wavelengths the effect of the interband transitions produces large differences. These decrease as the temperature increases, to such an extent that the temperature coefficient of the spectral emissivity passes through zero in a narrow wavelength region known unimaginatively as 'the X-point', and becomes negative at short wavelengths.

Early interest in the Drude theory was directed more towards the calculation of total emissivities for heat transfer applications. In 1905, Aschkinass integrated the Hagen-Rubens equation over all wavelengths to obtain the total normal emissivity. Foote later extended the calculation by including the second term of the expansion in (3.34). His result can be given in the form

$$\epsilon_N = 0.578(r_e T)^{\frac{1}{2}} - 0.178(r_e T) + 0.0584(r_e T)^{\frac{3}{2}} \tag{3.35}$$

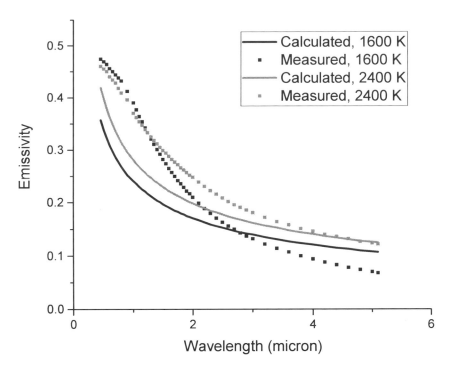

Figure 3.7: Hagen-Rubens calculation of ϵ_λ for tungsten compared to experimental data.

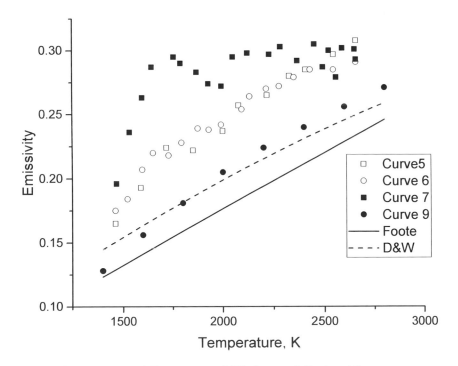

Figure 3.8: Comparison of Davisson and Weekes and Foote with ϵ_N measurements.

To keep with tradition, the resistivity r_e in this equation has units of $\Omega\,cm$.

In 1924, Davisson and Weeks employed the Fresnel relations to integrate the Hagen-Rubens equation over all angles of emission, calculating for the total hemispherical emissivity

$$\epsilon_T = 0.766(r_eT)^{\frac{1}{2}} - [0.0309 - 0.0889ln(r_eT)]R_eT - 0.0175(r_eT)^{\frac{3}{2}} \qquad (3.36)$$

A comparison of the measured values of tungsten ϵ_N with those predicted for ϵ_N and ϵ_T is given in Figure 3.8.

As the first terms in both (3.35) and (3.36) are the largest, and r_e for metals is generally approximately proportional to the absolute temperature, the emissivities increase linearly as the temperature is increased above ambient. Although it would be expected that deviations would be found when the fraction of thermal radiation at wavelengths below about $3\,\mu m$ to $5\,\mu m$ becomes significant, at about $600\,K$, in a few cases the dependence is linear up to much higher temperatures, for example around $2000\,K$ in platinum. At the time, these equations were regarded as successfully predicting the values of normal and hemispherical total emissivities for metals.

However, it is clear from Figure 3.8 that significant deviations from the predicted curve do exist, some of which cannot readily be explained. It may be noted that the neglect of other forms of electron scattering in the Drude theory will give calculated emissivities which are low, while imperfections in the measured surfaces will give experimental results which are higher than the ideal values. Neither can explain the existence of measured values below the predicted curve, as observed for copper, silver and gold. In fact it seems that any agreement between estimates derived from the Hagens-Rubens equation and experimental results is fortuitous. The fundamental assumption $\omega\tau_r \ll 1$ is invalid for metals at temperatures above $0\,°C$ and at the wavelengths of interest here, and the effect of this error is to increase calculated emissivities which should be low into rough agreement with the higher experimental results.

In fact, for the situations in which infrared radiation thermometry might be applied the opposite condition $\omega\tau_R \gg 1$ is more nearly correct, that is, metal surfaces fall in the Mott-Zener regime in which

$$\epsilon_{\lambda,N} = \frac{2}{\omega_p\tau_R} \qquad (3.37)$$

A simple interpretation would suggest therefore that the spectral emissivity is independent of wavelength in the mid-infrared. While this is the case for a few 'good' metals, gold for example, in most some allowance must be made for additional contributions to electron scattering rate which affect the relaxation time τ_R. At wavelengths for which the Mott-Zener condition does not apply with sufficient force, it is necessary – and not difficult for numerical calculations – to return to the basic Drude equations (3.29).

At temperatures above the Debye temperature of the metal the usual electron-phonon relaxation time in the bulk material may be written, from (3.30) and (3.33), in the simple form

$$\frac{1}{\tau_p} = \left(\frac{N_e e^2}{m_e} \right) r_e \tag{3.38}$$

It is also necessary to include a term to allow for the diffuse scattering of the electrons at the surface. The relaxation time τ_s is related to the transit time of the electrons across the skin depth

$$\frac{1}{\tau_s} = \frac{3\omega_p v_F}{8c_0} \tag{3.39}$$

where v_F is the Fermi velocity

$$v_F = \frac{h}{2\pi m_e}(3\pi^2 N_e)^{\frac{1}{3}} \tag{3.40}$$

The overall relaxation time τ_R, obtained from

$$\frac{1}{\tau_R} = \frac{1}{\tau_p} + \frac{1}{\tau_s} \tag{3.41}$$

which when substituted into (3.37), together with the plasma frequency from (3.30), gives an expression for the spectral emissivity which contains two parameters, r_e and N_e, both of which may be determined by electrical measurements. N_e may be obtained, for example, from the Hall constant

$$R = \frac{1}{eN_e c_0} \tag{3.42}$$

Apart from the small effects of thermal expansion, N_e should be independent of temperature in metals.

While the results obtained agree well with the experimental measurements of the emissivity of carefully prepared clean surfaces for some metals at wavelengths above 3 μm to 5 μm, significant deviations are found with others, particularly the transition metals, and at shorter wavelengths. The effects of electron-electron scattering may be allowed for by adding another term to (3.41):

$$\frac{1}{\tau_e} = K\omega^2 \tag{3.43}$$

Although the functional dependence of this relaxation time appears correct, the magnitude of the temperature-independent constant K required by the emissivity measurements is a factor of ten different from that derived from thermal and electrical estimates. Some compensation for the effect of interband transitions at the shorter wavelengths may be obtained by increasing the constant ζ_0 in the real part of the dielectric function in (3.29). The required value may be estimated from optical measurements of the surface at room temperature. Unfortunately, as may be seen from Figure 3.7, the interband contribution to the spectral emissivity appears to be noticeably temperature dependent so that the correction is of limited validity except at long wavelengths where it is small.

Finally, we may also note that the scattering at the surface may not be completely diffuse in nature, that is, some electrons may be specularly reflected and not lose momentum. For these metals, an adjustable parameter must be introduced into (3.39). The effect of this, the uncertain estimates for the constant K and the temperature dependence of the interband component means that the Drude free-electron theory cannot be employed to predict the emissivities of metals and alloys of unknown optical characteristics. The models briefly described above may be used to study and parametrise the variation of the measured emissivities with temperature and wavelength; the present level of knowledge of the optical behaviour of metals is insufficient for real applications in general.

Even if the required parameters for the calculations are available, the emissivities calculated are appropriate only for materials with smooth, clean surfaces. In the next section, complicating factors including the effects of surface roughness and of films will be considered.

3.3   COMPLICATING FACTORS

For many materials, particularly metals, the absorption coefficient is high. The penetration depth of light can be as little as a few nanometres at shorter wavelengths, and certainly of the order of $1\,\mu m$ for a typical radiation thermometer and metal. It follows that the optical properties are often dominated by the surface condition, including surface layers and roughness.

3.3.1   Surface layers

Oxide layers can be a particular problem with metal processing, causing large errors if the emissivity of unoxidised metal is used. Moreover, the layer thickness is liable to be temperature and time dependent causing unpredictable and significant changes during processing. In some cases, such as the semiconductor industry, layers can be a complicated and deliberate part of the material being measured.

The Fresnel equations (3.22) and (3.24) give the reflectivity for transparent and absorbing materials at normal incidence. The addition of layers adds to the complexity. To assess the effect on emissivity it is probably sufficient to consider a single layer. The reflectivity of a single absorbing layer on a substrate can be calculated [15]. This requires data on optical properties for values for the real and imaginary parts of the complex refractive index, such as may be found in [16].

Considering a substrate with index $\boldsymbol{n_2} = n_2 - ik_2$, with a single layer of index $\boldsymbol{n_1} = n_1 - ik_1$ and thickness d_1 in a non-absorbing atmosphere of index n_0, then for wavelength λ and at normal incidence the reflectivity R is given by [15]

$$R = \frac{t_{12}^2 + u_{12}^2}{p_{12}^2 + q_{12}^2} \qquad (3.44)$$

where

$$p_{12} = p_2 + g_1 t_2 - h_1 u_2$$
$$q_{12} = q_2 + h_1 t_2 + g_1 u_2$$
$$t_{12} = t_2 + g_1 p_2 - h_1 q_2$$
$$u_{12} = u_2 + h_1 p_2 + g_1 q_2$$

with

$$g_1 = \frac{n_0^2 - n_1^2 - k_1^2}{(n_0 + n_1)^2 + k_1^2} \qquad\qquad h_1 = \frac{2 n_0 k_1}{(n_0 + n_1)^2 + k_1^2}$$

$$g_2 = \frac{n_1^2 - n_2^2 + k_1^2 - k_2^2}{(n_1 + n_2)^2 + (k_1 + k_2)^2} \qquad\qquad h_2 = \frac{2(n_1 k_2 - n_2 k_1)}{(n_1 + n_2)^2 + (k_1 + k_2)^2}$$

$$p_2 = \exp^{a_1} \cos \gamma_1 \qquad\qquad q_2 = \exp^{a_1} \sin \gamma_1$$

$$t_2 = \exp^{-a_1}(g_2 \cos \gamma_1 + h_2 \sin \gamma_1)$$
$$u_2 = \exp^{-a_1}(h_2 \cos \gamma_1 - g_2 \sin \gamma_1)$$

and

$$a_1 = \frac{2\pi k_1 d_1}{\lambda} \qquad\qquad \gamma_1 = \frac{2\pi n_1 d_1}{\lambda} \text{(radians)}$$

3.3.2 Surface roughness

The Fresnel equations are appropriate where surfaces are fully specular. Real surfaces apart from liquids rarely are, and the surface condition will effect the emissivity. Wen and Mudawar [17] identify three different regimes depending on the root-mean-square roughness (σ, or r_a) to wavelength ratio.

If $\sigma/\lambda < 0.2$ then a simple analytical approach assuming a Gaussian distribution of surface height variation gives good agreement to experimental results. The reflectivity of a rough surface (ρ_r) is found to be reduced from that of the ideal smooth surface ρ_s according to

$$\rho_r = \rho_s \exp\left(\frac{4\pi\sigma}{\lambda}\right)^2 \qquad (3.45)$$

If $0.2 < \sigma/\lambda < 1.0$, then numerical methods and ray tracing are necessary to evaluate the bidirectional reflectance distribution function (BRDF) and so determine a value for ϵ. As the roughness of the surface increases beyond the wavelength of the thermometer the surface presents a series of facets, and the problem becomes a geometrical one. For example, concentric rings and pyramid structures can be used to enhance the normal emissivity of black-body cavities designed to be high emissivity sources.

One issue that needs to be considered at high temperatures is that heat treatment can modify the surface roughness, giving a permanent change to emissivity.

3.3.2.1 Angular dependence

The emissivity depends on the angle of viewing. It has already been noted that the Fresnel equations become more complex away from normal, and increasingly so if surface layers are present. In general the emissivity becomes decreasingly small at glancing angles, and this is also the case for rough surfaces.

3.3.2.2 Translucent materials

For translucent materials where k is small, the emissivity becomes a bulk property and is less affected by the surface condition. For example, in glass manufacture depending on the spectral region chosen the bulk temperature or surface temperature can be measured. If the sample is not thick enough to be optically opaque then the emissivity becomes a function of thickness.

CHAPTER 4

Radiation thermometer design considerations

4.1 CLASSIFICATION OF RADIATION THERMOMETER TYPES

Up to this point in the book, we have discussed the basic principles of radiation thermometry and the optical characteristics of surfaces. We now move on to a more practical topic, that of the design of radiation thermometers, and the choice of components available. Before we enter into the details, it may be worthwhile to present briefly an outline of the different types of radiation thermometers which the reader may come across. The basic value of a classification scheme, of course, is that it allows instruments with similar characteristics to be grouped together, and, conversely, that instruments which show significant differences in their essential properties to be separated. In seeking a radiation thermometer for a particular application, a sensible classification scheme will allow the range of suitable instruments to be narrowed down quickly and efficiently.

The choice of fundamental criteria for the separation into different groups is not as simple as might be thought. Some properties of radiation thermometers, such as the operating range of temperature, which are vitally important in selecting a thermometer for a given task, are unsuitable as a means of classification because they are continuous parameters. Not only is the division into, for example, low, medium and high temperatures a quite subjective decision which will vary with the scientist or engineer involved and their usual field of work, but it will also tend to be unstable, in that improvements in optical design and photodetector performance will extend the operating range of a given thermometer type to lower and lower temperatures.

There are in fact a number of supplementary points which may be of importance in the selection process, but which are not vital to the basic operating characteristics. Example are whether the instrument is portable, the nature of the display of the reading, the method of sighting on the target, and so on. The essential difference between the main groups of radiation thermometers is, in the author's opinion, in the selection of the wavelength range or ranges detected. This leads to the following classification scheme.

(a) Total radiation thermometers

This group accepts thermal radiation over a very wide range of wavelengths (in principle the complete spectrum), and detects it evenly. This means that neither the transmission of the optical system nor the response of the detector should be significantly wavelength dependent over the whole of the wavelength range emitted by the target. It is not a simple matter to satisfy these requirements, and genuine total radiation thermometers with useful sensitivity are found only in research laboratories. If the conditions are met, the instrument obeys the Stefan-Boltzmann law, and only a single calibration point is required.

(b) Broadband radiation thermometers

The signal from these is derived from a wide, more or less continuous[1], range of wavelengths, so that it is represented by the integral of the measurement equation over a broad band. As a result, these instruments do not follow, even approximately, either Planck's law or the Stefan-Boltzmann law, and they must be calibrated empirically against a standard source, preferably a black-body, at a number of temperatures. In use it is particularly difficult to allow for surface emissivities as this must be assessed over the complete waveband and varies from one application to another. These thermometers are typically of simple design aimed at industrial use where reliability and robustness are paramount, and either do not contain a colour filter to limit the optical bandwidth or may use simple coloured glasses with broad spectral transmission. Two main subgroups may be distinguished. The first contains those instruments with thermal detectors, whose spectral response is flat over a wide range. The wavelengths contributing to the signal are limited by the transmission of the optical system, including any windows used to protect internal components. The second group contains photon detectors, and these generally possess a well-defined cut-off which limits the response at long wavelengths. In both cases, the rapid decrease in the Planck distribution removes sensitivity to short wavelengths. A further sub-category is for measurement of lower temperatures in standards laboratories and national measurement laboratories, where broadband radiation thermometers will be required with precision as a priority, and these may be calibrated at a series of temperatures using fixed point cell black-body sources.

(c) Narrow-band radiation thermometers

In these, the optical bandwidth is limited by a colour filter with a relatively narrow pass-band. To overcome the consequent reduction in the amount of radiation reaching the detector, photon detectors, which are more sensitive than thermal detectors, are generally used. Occasionally the long wavelength cut-off

---

[1]Occasionally a 'notch' filter may be used to limit response to a problematic wavelength band such as water absorption bands in the mid-IR.

of the photon detector may be employed to limit the optical bandwidth. If the waveband is narrow enough and the detector is linear the output signal follows Planck's law. This group includes most of the instruments used as primary and secondary standards in calibrating laboratories.

(d) Multi-wavelength radiation thermometers
By measuring the signal levels over two or more wavelength bands, it is possible to compensate to some extent for two effects which can cause large errors with other types of radiation thermometers. The first is associated with small targets which do not fill the target area of the instrument, while the second is observed with surfaces with low or unstable emissivity characteristics. Given that significant errors and uncertainties arise from the effect of unknown emissivity it is tempting to suggest the problem can be alleviated by using two or more wavelengths. As will be shown later, there is never enough information to do this and additionally the more channels there are the more sensitive the method is to the assumptions made in processing the signals.

At this point we note that other criteria have been proposed for the classification of radiation thermometers. These include a division into direct reading instruments, in which the detector output is analysed to obtain the source temperature, and comparison types, which contain an internal reference source, against which the radiation from the source is compared. The requirements for stability and linearity of the detector are then much less stringent, as it forms part of a null detection system, and the calibration of the instrument is carried by the reference source, usually a small filament lamp or, at low target temperatures, a small electrically heated black-body. The division into visual, that is, those employing the eye as a detector, and objective is largely obsolete, as the disappearing filament pyrometer is virtually the only remaining member of the former class and itself rarely used.

4.2 THE GENERAL MEASUREMENT EQUATION

We now consider the effects of the main components of a radiation thermometer upon the signal obtained from the radiation detector. We shall take as an example a simple direct-reading instrument as shown in Figure 4.1. The three essential elements are the optical system, the colour filter and the detector. The signal amplification, processing and display components are of course important in practice but need not concern us at this point. The optical system serves to collect the thermal radiation from the source and direct it through the filter onto the detector. While it may include windows, baffles and other components which serve a range of functions, we are specifically interested here in those characteristics which determine the amount of thermal radiation transmitted.

We may write a general equation for the signal from the detector by considering the spectral radiance L_λ entering the thermometer through an element of area dA of the entrance aperture in a direction described by angles (θ, ϕ) to the optical axis.

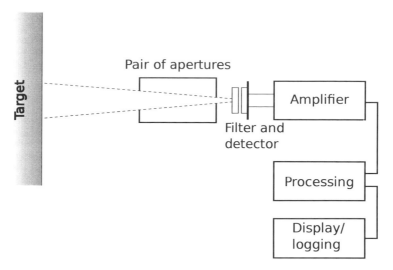

Figure 4.1: Simple direct-reading radiation thermometer.

This 'measurement equation' is of the form

$$I = \int_A \int_\omega \int_\lambda L_\lambda \cos\theta \, t_o(\lambda) \, t_f(\lambda) \, R(\lambda) \, d\lambda \, d\omega \, dA \qquad (4.1)$$

where $t_o(\lambda)$ and $t_f(\lambda)$ represent the spectral transmission functions of the optical system and the colour filter, respectively. $R(\lambda)$ is the spectral responsivity of the detector, expressed in terms of the output photocurrent or voltage per watt of incident radiation at each wavelength.

Apart from the $\cos(\theta)$ term required to give the transmitting area normal to the cone of acceptance of rays entering the thermometer (i.e. the field-of-view of the instrument), the directional dependence of the quantities in (4.1), that is, those that might depend on the trajectory of the ray through the instrument, has been suppressed. In general, the neglect of the directional quantities (θ, ϕ) may be justified on two grounds. First, the contribution of a particular ray to the output signal is largely independent of its trajectory, partly because the angular field of view is rarely large, but also because the directional dependencies of the transmission of the optical system and the response of the detector are in any case small. Although the characteristics of the colour filter may depend upon the angle of incidence, if the range of angles remains the same from one measurement to another, the effect upon the output signal will average out. In very accurate radiation thermometers, where this effect might lead to detectable errors, it may be reduced by collimation of the beam passing through the filter.

Secondly, we may note that, even if the output signal is to some extent dependent upon the way the radiation from the target area is distributed amongst the various optical paths through the instrument, the effect is independent of the temperature being measured and will cancel out for sources whose emission properties are independent of direction, that is, for all diffuse emitters. If this is not the case, and in particular we have shown in the previous chapter that it is not for real surfaces emitting at angles well away from normal, then a significant error may result. It may be remarked that the magnitudes of such errors have been rarely investigated.

With this reservation in mind, it follows that we may usually separate (4.1) into the form

$$I = \int_A \int_\omega \cos\theta \, d\omega \, dA \int_\lambda L_\lambda \, t_0(\lambda) \, t_f(\lambda) \, R(\lambda) \, d\lambda \qquad (4.2)$$

There are of course a number of small deviations from this equation, for example from diffraction effects, which must be dealt with by applying small corrections, determined theoretically or experimentally, but it provides a valuable tool for studying the performance of pyrometers in many applications. In fact there are two distinct ways in which it is applied in this book:

(a) In the design of a radiation thermometer to a given specification or for a particular application. In this case we need absolute values for the signal level I, to compare with predicted noise levels which may limit the accuracy of a measurement, and all the integrals in (4.2) must be evaluated. However, since there is often a wide variation in the responsivity and noise level obtained with a given type of photodetector, it is not necessary to do this with any great accuracy. In addition, it is often preferable to build to a conservative design with a safety factor of two or more in the estimated signal level. It follows that approximation methods may be generally applied and that small deviations from (4.2) are unimportant and may be neglected. The application of these methods will be described later in this chapter.

(b) To convert measured signal levels into a determination of the real temperature of a source, using Planck's radiation law with known emissivity values to obtain the spectral radiance, L_λ. In this case we may consider that the first term in (4.2), which is independent of the wavelength and source temperature, has been effectively determined experimentally during the calibration of the pyrometer, and does not need to be considered further. Similarly, only relative values are required for the terms $t_0(\lambda)$, $t_f(\lambda)$ and $R(\lambda)$ in the second term. However, if accurate values for the source temperature are required, the magnitude of the corrections for deviations in (4.2) will need to be considered.

4.2.1   The throughput of the optical system

The integrated product, G, of the projected area and solid angle

$$G = \int_A \int_\omega \cos\omega \, d\omega \, dA \qquad (4.3)$$

with units $m^2 \, sr$, is known variously as the throughput, the geometrical extent and the etendue of an optical system. For our purposes, we may estimate its value using simple geometrical optics, neglecting the effects found in real systems of scattering, diffraction and aberrations in optical components, none of which are important for the applications in which we shall need quantitative values of G. In addition, we note that the loss of signal from absorption in optical and unwanted reflections at their surfaces may be included in the term $t_o(\lambda)$. We first consider as a very simple example of an optical system a pair of circular apertures placed concentrically along an optical axis as depicted in Figure 4.2.

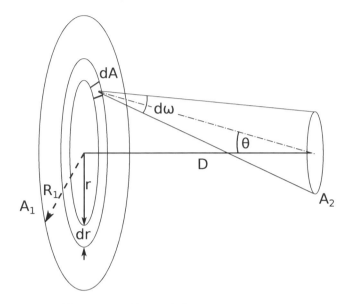

Figure 4.2: Beam-defining aperture pair.

In a real instrument, these would form the beam-defining aperture pair, that is, they alone determine the amount of radiation that is accepted from the source. All other components, baffles, lenses and so on, are assumed to be larger than the beam limited by this aperture pair in the plane in which it is normally located. This does not mean, of course, that the area of either of the apertures may not be determined, for example, by the clear area of a lens, or the active area of a photodetector.

We take the plane of the entrance aperture as the reference plane, and calculate the throughput of the transmitted beam from a source of infinite extent by integrating over each pencil of rays passing through an elemental area dA of the aperture, subtending a solid angle $d\omega$ at the second aperture. In general this calculation can involve some quite complicated geometry, which may require numerical techniques for its solution. The case presented here is a simple one capable of analytic solution, which nevertheless is not too dissimilar to many practical configurations, and which may be used to estimate the throughput G in real applications.

The beam-defining aperture pair consists of a large entrance aperture of radius R_1 and area A_1, separated by a distance D from a smaller exit aperture of area A_2. We consider first the contribution to the throughput from an annulus of radius r on the entrance aperture. The solid angle $d\omega$ subtended by the exit aperture from any point on the annulus is constant and given by

$$d\omega = \frac{A_2 \cos \theta}{(D^2 + r^2)} \qquad (4.4)$$

where θ is the angle between the axis and the line joining the annulus to the centre of the exit aperture. As the area of the annulus dA_1 is

$$dA_1 = 2\pi r \, dr \qquad (4.5)$$

the total throughput is

$$G = \int_A \int_\omega \cos \omega \, d\omega \, dA$$

$$= \int \cos \theta \frac{A_2 \cos \theta}{(D^2 + r^2)} 2\pi \, rdr$$

Noting that

$$r = D \tan \theta$$

$$dr = D \sec^2 \theta \, d\theta$$

substitution transforms the integral into the form

$$G = \int_0^{\theta_0} 2\pi A_2 \sin \theta \cos \theta \, d\theta$$

$$= \pi A_2 sin^2 \theta_0 \qquad (4.6)$$

where θ_0 is the maximum angle subtended by the entrance aperture from the centre of the exit aperture. By substituting for $\sin \theta$, we alternatively write the result in the form

$$G = \frac{A_1 A_2}{(D^2 + R_1^2)} \qquad (4.7)$$

It is sometimes stated that the throughput G for a given optical system is invariant. As G depends only upon the geometry of the system, this does not appear to add greatly to our understanding of the problem. However, if we include the spectral radiance, we obtain the result that, neglecting losses from absorption, scattering and so on, the total flux in the beam is independent of position along its length. Strictly speaking, if the beam passes through regions with varying refractive index n, it is the product $n^2 G$, known as the optical extent, which is invariant. It follows that the flux reaching the detector from a surface at uniform temperature does not depend upon the distance between the surface and the optical system of the pyrometer, as long as the surface is larger than the extent of the defined beam at its position, that is, it fills the optical system. Put another way, with this optical system, increasing the distance between the source and the instrument simply increases the target area but does not change the reading obtained if the target is uniform.

While this is a distinct advantage in the design of a radiation thermometer, an optical system consisting only of a pair of apertures is not suitable for most applications. While the entrance aperture must be reasonably large to admit sufficient flux, the exit aperture must be small to match the beam to the active area of the photodetector. As a result, the target area on the source is large for reasonable operating distances between the source and the pyrometer. Moreover, the target area is not well defined, as the contribution from regions around the periphery falls off gradually to zero. This effect, known as vignetting, is due to the fact that some rays from points in these regions directed toward the exit aperture are blocked by the entrance aperture, as shown in Figure 4.3.

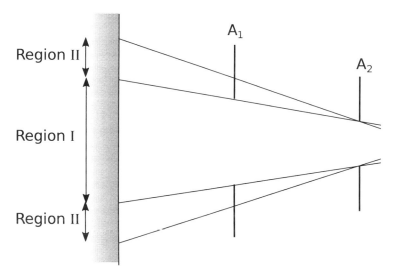

Figure 4.3: At a point in region I all the rays directed towards aperture A_2 also pass through the entrance aperture A_1. At points in region II some of the rays directed towards aperture A_2 are blocked by the entrance aperture.

Both difficulties may be largely removed by incorporating focusing elements within the optical system. A simple example is shown in Figure 4.4, with a convex lens placed between two apertures. It may be shown that this system is equivalent to a beam-defining aperture pair with no focusing elements if one of the apertures is replaced by its virtual image in the focusing system. Clearly the lens aperture must be larger than that of the entrance aperture, or it would itself become one element of the beam-defining pair. The entrance aperture is in this case more commonly called the aperture stop, and its function is to limit or define the solid angle of rays accepted from the source by the instrument.

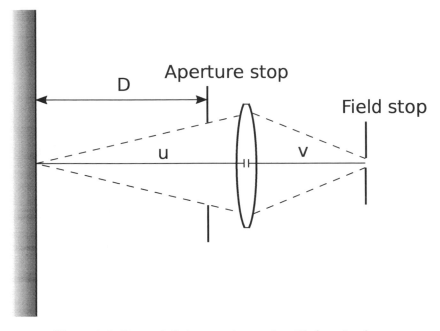

Figure 4.4: Beam-defining aperture pair with focusing lens.

In normal operation, the source is focused by the lens onto the exit aperture. It follows that the second element of the equivalent beam-defining aperture pair, the virtual image of the exit aperture produced by the lens, is formed by the target area. For this reason, the exit aperture is more usually known as the field stop. It will be seen that this arrangement is particularly simple to analyse; the throughput is given by the product of the target area and the solid angle subtended by the aperture stop, which is sometimes known as the angular field of view.

We might ask whether the measured signal remains constant as the lens is moved to change the operating distance. From Figure 4.4, the target area is $A_2(u/v)^2$ where A_2 is the area of the field stop. The solid angle subtended by the aperture stop is $A_1(v/D)^2/u^2$, so that the total throughput is A_1A_2/D^2, that is, independent of the target distance as long as the aperture separation is not changed.

The main advantage in the use of a focusing system is, of course, that the target area is many times smaller than that realised with a simple aperture pair. It is also well defined in that, neglecting diffraction effects and lens aberrations, the contribution to the output signal is uniform across the whole of the target area. The good definition of the beam also reduces the chance that stray radiation may enter the radiation thermometer. These advantages are obtained without a significant reduction in the throughput, although some radiation is of course lost by absorption in the optical components, and by scattering and diffraction at their surfaces.

In general, the throughput of an optical system is not required with great accuracy, and a number of options are available to obtain estimates of this quantity. Equation (4.7) may be employed if the assumptions associated with its derivation are reasonably well satisfied. Alternatively, (4.3) may be solved by numerical integration over the effective beam-defining pair. Finally, if all else fails, it may be measured experimentally. This approach is not recommended, as it is possible to introduce significant errors if the requisite care is not taken.

4.2.2 Noise-limited performance

Ultimately, the performance of any radiation thermometer is limited by the level of noise in the measured signal. We are of course using the word 'performance' in a very specific way in this context. It does not include factors which characterise the source whose temperature is being measured, but is intended to indicate the optimum behaviour of the radiation thermometer under ideal conditions. We may assume, for example, that the source is a black-body at a stable temperature, and whose emitting area is considerably larger than the target area of the thermometer. Any quantities which might influence the accuracy obtained are taken to be set to their ideal or nominal values – the ambient temperature, for example. And the instrument is assumed to be correctly aligned and calibrated. Many of these factors may in practice contribute towards the final uncertainty in an actual measurement. What we will attempt to calculate in the following section is the lowest temperature at which the desired resolution can be obtained in a typical measurement period. The operating range can always be extended to higher temperatures by inserting calibrated filters into the beam to reduce the signal level to within the working range of the detector, but the noise in the output signal forms a fundamental limitation to the lowest source temperature which can be satisfactorily measured.

We shall now show how to relate the limiting temperature resolution to the characteristics of the components forming the instrument. The calculation will be carried out for a narrow-band radiation thermometer, as this forms a typical and useful example, and enables the integral over wavelength in (4.2) to be replaced by a simple product evaluated at the operating wavelength of the instrument. To extend the calculation to broadband radiation thermometers, the integrals in (4.2) must be estimated in some fashion, and this makes the procedure rather more cumbersome, but not greatly more complicated. Total radiation thermometers are a particularly simple case, of course, because the detection system must be insensitive to the wavelength of the thermal radiation, and the Stefan-Boltzmann law may be applied.

Taking the terms within the wavelength integral in order, we note first that we can use the Wien approximation for the spectral radiance. The effective wavelength of narrow-band radiation thermometers is usually on the low-wavelength side of the Planck distribution, where this approximation is valid, and almost certainly so at the low temperature end of the working range. Although we intend to carry out the calculation for a black-body source, if the pyrometer is being designed for a particular application where the spectral emissivity is known, this should be included at this point. Both the transmission of the optical system and the detector responsivity are taken as constants, and represented by t_o and R, respectively. The remaining term, the integrated transmission of the colour filter, is replaced by

$$\int t_f(\lambda)\,d\lambda = a_0\,t_f\,\Delta\lambda$$

where t_f is the peak transmission of the filter, $\Delta\lambda$ the full bandwidth measured between the half-intensity points (that is, the wavelengths at which the transmission has fallen to one-half its maximum value), and a_0 is a factor which describes the shape of the transmission curve $t_f(\lambda)$ and is usually about unity. Overall then, (4.2) is reduced to the form

$$I = G\,t_o\,R_0\,a_0\,t_f\,\Delta\lambda\,c_1\,\lambda_0^{-5}\exp(-c_2/\lambda_0\,T) \tag{4.8}$$

where λ_0 is the operating wavelength of the pyrometer and R_0 is the detector responsivity at that wavelength.

We may relate fluctuations ΔI in the signal, whether from noise, measurement errors or whatever, to the resulting uncertainty ΔT in the temperature by differentiating (4.8) to give

$$\frac{\Delta I}{I} = \left(\frac{c_2}{\lambda_0\,T}\right)\frac{\Delta T}{T} \tag{4.9}$$

Here we will consider only the effects of random noise in the output signal, in which case $\Delta I/I$ is the inverse of the signal-to-noise ratio (S/N). This is expressed in terms of the measured variable, either current or voltage. However, the noise may arise from a variety of sources, and, if they are independent, the total noise power is simply the sum of the noise powers from the individual sources. It is therefore convenient to work in terms of the ratio of the signal power to the noise power, $(S/N)_P$, where

$$(S/N)_P = (S/N)^2 \tag{4.10}$$

We will illustrate the calculation of $(S/N)_P$ for a simple photodetector which generates an output current i. The quantum efficiency μ is the probability that a photon will release a single electron to form part of this current. It is of course likely to vary with wavelength. If a monochromatic flux ϕ is incident upon the active area of the photodetector, then, since each photon has energy $h\nu$ the photon flux is $\phi/h\nu$ per second. It follows that the average number $\overline{n}$ of electrons generated in a measurement time τ is given by

$$\overline{n} = \frac{\mu\phi\tau}{h\nu} \tag{4.11}$$

We specify the average value because the detection process is random in nature, and the number n of electrons generated will vary from one measurement period to the next. The process obeys Poisson statistics, and as such it may be shown that the mean square fluctuation in n is equal to the mean value $\overline{n}$, that is,

$$<(\mathbf{n} - \overline{n})^2> = \overline{n}$$

The variable measured is usually the photocurrent i which is related to n by

$$i = \frac{en}{\tau} \tag{4.12}$$

where e is the charge on the electron. The mean photocurrent is given from (4.11) above by

$$\bar{i} = \frac{e\mu\phi}{h\nu} \tag{4.13}$$

and the mean square current fluctuation

$$\bar{i}_N^2 = <(\mathbf{i} - \bar{i})^2>$$
$$= \left(\frac{e}{\tau}\right)^2 <(\mathbf{n} - \bar{n})^2>$$
$$= \left(\frac{e}{\tau}\right)^2 \bar{n}$$
$$= \left(\frac{e}{\tau}\right)\bar{i}$$

More generally, the noise may be related to the bandwidth B of the detection system rather than the measurement time. The two are related for an integrating filter by

$$B = \frac{1}{2\tau}$$

and hence

$$\bar{i}_N^2 = 2e\bar{i}B \tag{4.14}$$

which is the well-known expression for shot noise, originally developed to explain the noise generated in vacuum diodes.

It is important to note that shot noise is generated by the total current from the photodetector, and this may contain components other than the required signal current i_S. In a more general (but still simplified) case, we may write for the detector current

$$\bar{i} = \frac{e\mu(\phi_S + \phi_B)}{h\nu} + i_D \tag{4.15}$$

where ϕ_S is the flux incident on the photodetector from the target area, ϕ_B the background flux from other sources, and i_D the dark current, that is, the current generated by the detector in the absence of incident radiation. It may be assumed that the undesired contributions to the detector current may be subtracted in the measurement of the signal current,

$$\overline{i_S} = \frac{e\mu\phi_S}{h\nu} \tag{4.16}$$

but of course the contributions of these to the mean square fluctuations cannot be removed. The total detector noise is in this case therefore given by

$$\overline{i_N^2} = 2eB\left[e\mu\left(\phi_S + \phi_B\right) + i_D\right] \qquad (4.17)$$

The background flux ϕ_B may come from stray radiation reflected from external sources into the instrument, or from thermal radiation generated by parts of the instrument itself. The effect is important only if the temperature of the sources or of the thermometer components is comparable to or greater than that of the target area. Although this additional signal is usually represented by an increase ϕ_B in the flux reaching the detector, it should be noted that the spectral distribution of the background radiation may be very different from that from the target area. The difference is limited by the transmission band of the colour filter. If the temperature of the source is low, around ambient, it is of course essential that the detector should not be able to 'see' thermal radiation from surfaces at ill defined or unstable temperatures and which has not passed through the colour filter. It is common to cool not only the detector and its immediate surroundings including any filter, but also stops and baffles which limit the passage of stray radiation to the detector.

In addition to shot noise there is thermally generated noise even when no current is produced. This results from the random motion of thermally excited electrons. The mean voltage variance in a frequency interval df due to the thermally excited electrons in a resistance R at a temperature T is given by

$$\overline{v}^2 df = 4k_B T R df$$

where k_B is the Boltzmann constant, and so the current fluctuation in bandwidth B is

$$\overline{i} = \sqrt{\frac{4k_B T B}{R}} \qquad (4.18)$$

In many cases the signal from the photodetector is too small, or at an inconvenient impedance level, for any useful purpose, and it is necessary to add some stage of amplification and current to voltage conversion. This also contributes to the level of noise in the output signal. The contribution is commonly expressed in terms of noise figure F or the related noise temperature T_N of the amplifier. We consider the amplification of a current source with an effective load resistor R. The noise figure F is defined as the ratio of the input signal-to-noise ratio to that at the output, with the load resistor at some reference temperature, usually 290 K. Since the signal is increased by the power gain A of the amplifier we have $F = (1/A)(N_{out}/N_{in})$ and so

$$F = \left(\frac{1}{A}\right)\left(A4k_B T_0 B + \frac{P_{ex}}{4k_B T_0 B}\right)$$

where $k_B T_0 B$ is the thermal noise from the load resistor, and P_{ex} is the excess power generated by the amplifier, that is, the noise added over and above the amplified input noise. Rearranging this equation yields

$$P_{ex} = (F - 1)k_B T_0 B\, G$$

In other words, the noise added by the amplifier can be treated as equivalent to an additional thermal noise component at the input. The noise temperature T_N is the temperature required to generate a power $P_N = k_B B T$, the equivalent output noise power due to the various circuit components in the amplifier. The thermal noise from R and the amplifier circuit add so that

$$T_N = T_R + (F - 1)T_0$$

with the noise current

$$i_N^2 = \frac{k_B T_N B}{R}$$

Adding this term to (4.15) gives the equivalent total noise power of the detector. Squaring (4.12) gives the signal power and so

$$\left(\frac{S}{N}\right)_{out} = \left(\frac{e\mu\phi_S}{h\nu}\right)^2 \left(2eB\left[\frac{e\mu(\phi_S + \phi_B)}{h\nu} + i_D\right] + 4k_B T_N B/R\right)^{-1} \tag{4.19}$$

To obtain $\Delta I/I$ in order to solve (4.9) we must invert this expression and take the square root. This will allow the smallest temperature difference for a given source temperature to be found, i.e., the instrument's resolution.

This assumes a constant noise factor for the amplifier. This is generally a reasonable assumption, but considering the common trans-impedance amplifier, at high frequencies the capacitive nature of a photodiode presents an alternative path to earth and the noise gain due to the op-amp is increased. The transimpedance amplifier (Figure 4.5) acts as a current-to-voltage converter and has unity gain $A = 1$: the op-amp holds both inputs at earth, so all the current flows through the virtual earth at point E. At high frequencies some current can flow directly to earth, so the gain drops. The internally generated noise, however, is increased by the non-inverting gain $A = (1 + Z_F/Z_s)$.

At d.c. Z_s is dominated by the capacitive behaviour of the photodiode (the leakage current to earth is very small) so the resistance (d.c. impedance) is very large. As the feedback impedance Z_F is dominated by the value of R_F, for any likely value of feedback resistor A is close to unity and so noise is not amplified. For a.c. systems, however, the photodiode capacitance allows conduction in the reverse direction, so Z_s is reduced at higher frequencies. This causes an amplification of the noise current, and also can cause oscillation in the circuit. A shunt capacitor across the feedback resistor equal to the capacitance of the photodiode will prevent the noise current being amplified, but will slow the response time. Reducing the feedback capacitance will allow higher frequency measurement, but at the expense of increased noise. The amplifier will start to 'roll off' at higher frequencies, but the increase in noise will happen before this. It may be an issue when very high speed measurements are needed.

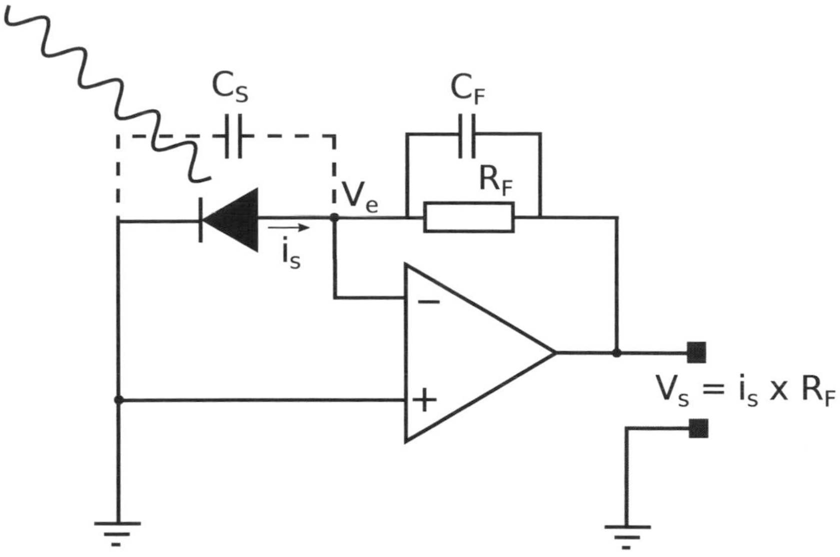

Figure 4.5: Photodiode with transimpedence amplifier.

4.2.3 Noise limits and examples of behaviour

From (4.19), it may be seen that there are four limiting cases, that is, where one noise source is predominant:

(a) At high signal levels, or if the detector noise level is low, the system is said to be signal or photon noise limited, that is, the noise arises from the fluctuations in the signal itself. While this represents the ideal case for photon detection, in practice the noise level for small signals is usually determined by one of the other noise sources. For the simple detector described above

$$\left(\frac{S}{N}\right)_{PL} = \frac{\mu\phi_S}{2h\nu B} \tag{4.20}$$

(b) At low target temperatures, the noise generated from extraneous radiation may predominate. In this case, the detection is background noise limited, and the signal-to-noise power ratio is given by

$$\left(\frac{S}{N}\right)_{BL} = \frac{\mu\phi_S^2}{2\phi_B h\nu B} \tag{4.21}$$

(c) If the total noise is determined by the dark current generated in the detector, the process is dark current limited, and

$$\left(\frac{S}{N}\right)_{DL} = \frac{e(\mu\phi_S/h\nu)^2}{2i_D B} \tag{4.22}$$

(d) Finally, if the contribution from the photodetector amplifier is the major source of noise, the detection process is said to be amplifier noise limited, with

$$\left(\frac{S}{N}\right)_{AL} = (e\mu\phi_S/h\nu)^2(R/4kT_N B) \tag{4.23}$$

At low signal levels, where the noise level is critical to the performance of the radiation thermometer, we usually find that one or other of these cases holds. We should stress that the discussion and equations given above only apply for detectors with simple characteristics, and we will find it necessary in the detailed treatment of some photo-detectors to modify some of the terms to take account of their real behaviour. In particular, it will be necessary to introduce the concept of detector gain where each photon detected does not produce exactly one electron, and then additional noise may be produced by the statistical fluctuations in the amplitudes of the current pulses. This is especially the case for photomultiplier tubes and avalanche photodiodes.

It is convenient to define, as a measure of the performance of a detection system, the noise-equivalent-power, or NEP, as the signal flux required to give a signal-to-noise ratio of unity. In the case of the simple detector covered above, and assuming that at low signal levels the photon noise contribution may be neglected, as is usually the case, then combining this definition with (4.19),

$$\left(\frac{e\mu NEP}{h\nu}\right)^2 = 2eB\left[\frac{e\mu(\phi_S+\phi_B)}{h\nu} + i_D\right] + \frac{4k_B T_N B}{R} \tag{4.24}$$

The right-hand side of this expression is part of the denominator in (4.19), and substituting and rearranging gives

$$\left(\frac{S}{N}\right)_{P_{out}} = \left[\frac{NEP^2}{\phi_S} + \frac{2h\nu B}{\mu\phi_S}\right]^{-1} \tag{4.25}$$

and hence

$$\left(\frac{\Delta I}{I}\right) = \left[\frac{NEP^2}{\phi_S} + \frac{2h\nu B}{\mu\phi_S}\right]^{1/2} \tag{4.26}$$

where the first term on the right-hand side represents the noise from the detection system, and the second that from fluctuations in the signal.

The NEP is therefore a very simple and convenient figure-of-merit for characterising a particular detection system at low signal levels, as it avoids the need to consider in detail the individual contributions to the noise. A more general figure-of-merit, the specific detectivity or $D*$, is widely employed to characterise a particular detector material or type. It is defined as the inverse of the NEP, normalised to a detector area of $1\,\mathrm{cm}^2$ and a detection bandwidth B,

$$D^* = \frac{AB}{NEP}$$

where A is the active area of the detector in cm^2. Clearly, the larger the value of $D*$, the more sensitive the detector is likely to be. It is necessary to specify the conditions under which any given value applies, for example, the wavelength of the radiation (usually that at which $D*$ is a maximum), the chopping or modulation frequency if it has not been measured under d.c. conditions, and the temperature of the source of radiation. Values of $D*$ are provided by the manufacturers of photodetectors, or may be found in the literature. For their application to the estimation of pyrometer characteristics to be valid, the conditions under which they have been determined should match reasonably closely those assumed for the pyrometer. If they do not, an adjustment, for example for the change in responsivity with wavelength, may be made. Alternatively, it may be possible to return to the detailed equation of the form (4.19) to obtain a closer approximation, if the detector has been well characterised and its operating parameters are known. The use of the parameter $D*$ is of course only valid if the reasoning behind it is correct, that is, that the mean-square noise fluctuations are proportional to the area of the detector. This is not always the case.

4.2.4   Additional noise sources

The above components – the shot noise from the signal itself and also from any background, thermal noise and any contribution from the current to voltage conversion and amplification – represent the lowest noise that can be achieved with a system. However, other effects can cause disturbance of electrons in the detector circuit and so contribute noise. For this reason it is best to keep connections as short as possible and to isolate the detection system from outside influences.

4.2.4.1   Mechanical noise

Simple mechanical action of conductors (wires) moving against the insulating sleeving can displace electrons and so give rise to triboelectric noise. Straining wires can introduce piezoelectric effects. Ideally wires are isolated from vibration but since this is not always practical low noise cables, which are readily available and designed to minimise this effect, should be used for the best measurement. Vibration can affect capacitors and resistors, giving audiophononic noise if the vibration is at audible frequency, so care should be taken in anti-vibration mounting and selection of suitable components in the signal path for any amplifier.

4.2.4.2   Electromagnetic interference

Any time-varying electric field can couple to the detection system if the system is not electrically isolated in a Faraday cage. It is therefore important that components are in a conducting enclosure connected to earth. The degree of shielding will increase with the thickness and conductivity of the enclosure. Any gaps or holes will decrease the effectiveness; conducting gasket materials are available to ensure a sealed enclosure. Any wire that feeds in from outside has the potential to act as an aerial and pick up interference, and this should be borne in mind. Use of twisted pair wiring or shielded, coaxial cables are standard practice. It is also important to avoid ground loops – the detection system and shielding should have a single, good earth connection. Note that this includes any voltmeter used to measure the output signal.

4.2.5   a.c. effects

Insulation is not ideal and small amounts of electric charge will be stored when a current first appears in a wire. At high frequency this can have an effect on the signal as it acts as a capacitor, and so low capacitance cable should be used if this is likely to be an issue.

4.3   RADIATION THERMOMETER DESIGN PROCESS

4.3.1   General radiation thermometer design considerations

In this section we consider a general procedure which may be adopted for the first stages in the design of a radiation thermometer. It will be assumed that the designer has a range of applications for the proposed instrument in mind, as it is these that will largely determine the essential characteristics. It should be noted whether these involve particular properties which might limit the range of options available. For example, an instrument which is required to be rugged and portable will probably not be able to use a photomultiplier as a detector, because they are fragile, rather large in size, and require a high voltage power supply.

The procedure to be described attempts to discover a set of parameters which satisfy (4.8) and (4.9) at the lowest operating temperature required in any of the applications. It is assumed that at this temperature the measurement repeatability is limited by the noise of the detection system, and that this forms a significant fraction of the overall uncertainty of measurement. Additional components, from calibration errors, the effect of ambient temperature fluctuations, the ageing of components and so on, must of course be considered and combined with the noise-limited repeatability to obtain the overall uncertainty, but these do not increase with falling temperature as rapidly as the repeatability. This feature is largely peculiar to radiation thermometry, and is a result of the rapid increase in radiance and hence signal level with increasing source temperature.

When the required repeatability and resolution have been obtained, it is of course important to ensure that the overall accuracy is not degraded by the sources of systematic errors at any temperature within the working range. However, whereas the noise-limited performance represents a fundamental limit for a particular radiation thermometer design, these other factors may usually be reduced by improved techniques or the selection of better, more stable, components.

The procedure involves 'cut and try' methods rather than a straightforward calculation of the required parameters. The major design choice is of course that of the detector; if it is found that the radiation thermometer cannot achieve the required noise-limited performance at the lowest operating temperature, with a given detection system and selected values for the physical parameters, then the latter may be varied in turn within ranges limited by cost and availability. If this fails, then it will be necessary to investigate another detector with greater sensitivity or capable of operating at a longer wavelength.

If successful, the results represent only a first stage in the complete design. For example, we may obtain a reasonable value for the throughput of the optical system, but this does not in itself decide if reflecting or refracting optics are to be used for the focusing elements. The procedure will yield a value for the operating wavelength, and this, together with the cost of the components required, is likely to determine the choice in this matter. There are also of course many other design points which are largely incidental at this stage; these include, for example, the form of the output signal, and the provision of alarm levels. These will be determined by the application range, and sometimes by the whim of the designer, but they do not affect the major choices which concern us here. In fact, radiation thermometers can be relatively complex instruments, and there are many more varieties of design than is the case with other thermometers such as platinum resistance thermometers or thermocouples.

We start, therefore, by determining the noise-limited repeatability δT required at the lowest operating temperature T_m. Conservatively, this may be taken to be one-half to one-third of the overall uncertainty required by the projected applications at this temperature. Next, the measurement time or electrical bandwidth is specified. If the object whose temperature is to be measured is moving rapidly, or if the apparent temperature is fluctuating or changing rapidly, a short measurement time may be necessary in order to resolve these time-dependent effects. This may restrict the range of suitable detectors. Thermal detectors, in particular, tend to have rather long response times, of the order of 1 ms to 100 ms, and may therefore be inadequate for these applications. If no such problems exist, a measurement time of between 0.1 and 1 second may be assumed. If it becomes necessary, a sequence of readings may be taken to reduce the spread. Applying statistics to the set often reveals useful information on the quality of the results.

The next step is to calculate an initial value for the throughput G of the optical system. This depends particularly upon two factors, the magnitude of the target area on the source, and the numerical aperture of the focusing elements in the optical system. Clearly, large optical components of high quality tend to be expensive, so it is best to adopt rather conservative values for the first approximation. The allowed range of values for the target area depends upon the magnification of the optical system, as explained in §4.2.1, and the sensitive area of the photodetector, as the image of the target area should not fill this. If only small detectors are available, the image can be reduced in size with additional optical components, but this is best avoided if possible, as it tends to make the thermometer more sensitive to vibration, and every additional optical component adds to the scattered radiation. In general, small target areas can only be achieved with focusing elements of short focal length, and with short distances between the source and the front element of the thermometer. Having assumed values for these quantities, the thoughput G may be estimated with (4.6) or (4.7).

4.3.2   Choice of wavelength and design procedure

At this stage we require a rough estimate of the operating wavelength. In some applications this may be decided by the properties of the material or system being measured or controlled. For example, many plastics have a strong absorption band at a wavelength $3.43\,\mu m$, corresponding to the stretching frequency of the C–H bond. This means that quite thin films appear 'black', that is, behave as a black-body, at this wavelength, whereas at visible wavelengths they are quite transparent. In others, it may be necessary to avoid wavelengths which correspond to absorption or emission bands in the medium between the source and the thermometer, for example the water vapour absorption bands in air, especially if the path length is large, and the various emission bands from flames which may be heating the furnace or the source itself.

If none of these limitations are relevant, then we may base our choice upon two conflicting requirements. At short wavelengths, we see from (4.9) that we obtain the lowest uncertainty in temperature for a given level of measurement error. Moreover, the cost and quality of optical components is at an optimum in the visible region of the spectrum. However, the spectral radiance from the source will increase rapidly with wavelength up to the maximum in the Planck distribution. As a rough guide, we should first consider detectors whose long wavelength cut-off is on the short wavelength side of the Planck distribution for the lowest operating temperature, i.e.,

$$\lambda_c T_m \leq 3000\,\mu m\,K$$

Having selected a rough operating wavelength, we may estimate the transmission of the optical system, including any additional elements and windows. Table 4.1 contains a list of the more common transparent or semi-transparent materials used for lenses and windows. From this, we may estimate the reflective losses at the optical surfaces and the absorption in the components having taken account of the possible effect of anti-reflection coating. We stress that, at this stage, approximate values are quite sufficient. We will assume at this point a rather narrow optical bandwidth, for example 2 % of the operating wavelength. The main advantage in doing this is that both the characteristics of the thermometer and of the surface being measured may often be known quite accurately in this case. If the detector is approximately linear, the output signal will follow Planck's law quite closely as the source temperature varies. If the resulting design lacks sensitivity but is otherwise acceptable, the bandwidth may be increased. An estimate of the colour or interference filter peak transmission may be made from the bandwidth and the centre wavelength. If low cost and simplicity of design are regarded as essential factors, it may be decided at this point to aim for a broadband instrument with no filter, or possibly an inexpensive coloured glass filter, at the expense of overall measurement accuracy.

Finally, we must select a photodetector. The various options are described in detail in a later chapter. Briefly, we note that response at wavelengths longer than the required operating wavelength effectively only increases the noise level. Thermal detectors are therefore only suitable when their low sensitivity (and the long response times) are acceptable, and usually for broadband designs. The responsivity of many photon detectors increases to a broad maximum with wavelength, and then drops off rapidly. Suitable detectors of this type should match the required operating wavelength to the maximum responsivity.

Parameters describing the detailed noise performance of a given detector are rarely fully available. It is usual, however, for manufacturers to give values for $D*$ and the NEP. While the noise level may be reduced, in some cases dramatically, by cooling the detector and its associated pre-amplifier, this adds considerably to the complexity of the instrument and its mode of operation. As the surface temperature being measured decreases to ambient and below, it becomes of course increasingly beneficial to cool the detector and also elements in its field of view. Probably the best compromise are then the small Peltier-cooled devices, which can be readily stabilised at moderately low temperatures. Finally, we note that if a range of detectors is available with reasonable sensitivity at the wavelength required, then the first choice would be those which are known to be stable and to have nearly linear response. In addition, some detectors may be fabricated in a range of sizes. While detectors with very small active areas may give rise to problems with alignment, those with large active areas have slower response times and increased noise levels. Clearly, if a range of sizes is readily available, it is easier to select a detector type whose performance represents a reasonable compromise.

Together with (4.26) we now have enough information to solve (4.8) and (4.9). Although we have stressed the significance of the minimum operating temperature, it is a simple matter to generate a set of curves for a given detector which indicates the noise-limited resolution T_{NL} to be expected as a function of the source temperature T_S for a number of wavelengths around the estimated operating wavelength, and the choice of wavelength may be optimised in this way.

We note in passing that if the resolution is everywhere much poorer than that required, this may indicate that the operating wavelength is too high, and it is necessary to consider whether the choice of a detector responding to shorter wavelengths might not result in an improved design. If, however, it is impossible to find a wavelength at which the noise-limited resolution meets the requirement at the lowest operating temperature, then it may be possible to achieve an acceptable solution by modifying the estimates of the other parameters or the operating conditions. We may consider in turn increasing the optical bandwidth or the throughput of the optical system, or cooling the detector. Each of these has disadvantages which have been discussed in the preceding section, but which might be considered acceptable if no alternative exists. If they are not, then it is necessary to select another detector and repeat the whole procedure, or else accept a compromise on the specification.

When a satisfactory set of parameters is found, then the second major stage of the design procedure is entered. In this, it is necessary to determine the 'finer details'. In particular, we need to estimate the long-term stability of the instrument, and the effect of ambient temperature changes. These will depend upon the characteristics of the major components. In some cases this information will be available from the manufacturer or in the technical literature. If it is not, then tests should be set up to measure the required quantities. The maximum allowable drift rate will of course depend upon the uncertainty required and the frequency of re-calibration. The ambient temperature coefficient, expressed as a change in the reading for each degree change in the ambient temperature, will be limited by the environment in which the instrument is to be used. Clearly, the errors produced in a standards laboratory, where the ambient temperature is tightly controlled, will likely be much less than in a factory or similar environment.

The stability and linearity of the detector also determines whether the thermometer may be a direct reading instrument, or whether it must include a reference source so that null detection techniques can be used. An internal reference source also greatly reduces the demands of stability and low temperature coefficient on all the other components, as its presence effectively continuously calibrates the system. The characteristics desired are then of course transferred to the reference source, which must demonstrate both short- and long-term stability.

In addition, there are many minor factors which can influence the final design. The portability of the instrument, the display of the temperature reading and the provision of alarm signals are only a few of these. And overall, the question of cost may influence almost every design choice, and it is essential that the role of the instrument in the marketplace should be well established at an early stage. It should be noted that the cost of the components and that of manufacture form only part of the final price. Radiation thermometry is a much more complex subject than other areas of thermometry, and the manufacturer may be expected to provide information and expertise on the usage of their instruments in the applications for which they were designed and sold.

4.4   OPTICAL SYSTEM DESIGN

The overall layout will be strongly influenced by, firstly, the wavelength range required and, secondly, the degree to which optimum optical performance is required. The choice of wavelength, or wavelengths, will determine which optical materials can be used, and whether a reflecting or a refracting design is most suitable. So, for example, as we will see below, the best optical performance will reduce uncertainties, but will make the final instrument longer, bulkier and more expensive than otherwise. Some designs briefly considered here are divided into reflecting and refracting optics

4.4.1 Refracting optics

4.4.1.1 General principle

The target size and aperture stop size will already have been determined by the required throughput. With single lenses a symmetrical layout will have the best optical imaging, but such a layout is often impractical as it makes for a large instrument and a short distance between the target and the first optic. This approach might be taken when, for example, making radiometric measurements of radiance temperature as it minimises the number of surfaces and absorbing materials, and consequently simplifies the calculation of G with reduction in this component of uncertainty. However, this is usually an inappropriate design for a radiation thermometer. There is an advantage to using achromatic lenses even for single wavelength narrow-band instruments. Achromatic lenses using two different glasses are typically designed for use at infinite conjugate ratio. A pair of lenses can be used back-to-back, but this increases scatter and ghost images, and may require additional stops to reduce stray light. In practice, standard achromatic lenses are expected to have good performance to at least 5:1 imaging, forming an image of the target at the field stop one-fifth the size of the field-of-view. Lenses should be installed so the surfaces spread the work of focusing – typically this means the most curved surface facing to the target. Since the choice of infrared materials is limited the use of achromatic lenses for longer wavelengths may not be possible.

4.4.1.2 Refracting optics – simple design

The simplest system is, as shown above, a single lens with two apertures. The advantage of this design is cost and simplicity, and in circumstances where the lowest uncertainties are not needed, or are dominated by other processes, this makes a perfectly sensible design choice. It is also a design where the throughput can be calculated and so can be the basis of an instrument for absolute radiance measurements.

4.4.1.3 Refracting optics – Lyot stop design

Rejection of scattered and stray light is a primary requirement for the most accurate radiation thermometry. The Lyot stop, named after French astronomer Bernard Lyot, was first used in 1931 for a coronograph, where the principle requirement is to block the image of the sun while still being able to view the much fainter corona. A coronagraph has one or more stops, each of which is in the image-plane of an aperture stop that will be scattering light towards the detector. The size of the Lyot stop is such that its magnified image matches the size of the actual aperture. Only light that makes up the optical throughput at the first stop is transmitted through the Lyot stop and on to the detector. Each potential scattering aperture needs a corresponding Lyot stop. The position of the stops can be calculated [18]. Figure 4.6 shows a design for a refracting system that has very low scatter. It can be made such that the only stray light contribution to the signal is the result of flare

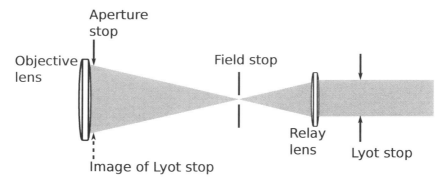

Figure 4.6: In the Lyot stop design the relay lens forms an image of the Lyot stop at the aperture stop.

in the objective lens. It may therefore be the dominant contributor to light from outside the nominal field-of-view that gives rise to the size-of-source effect where the measured signal is different for targets of different size. The disadvantage of this system is that it requires an extra lens and sufficient distance to position the Lyot stop, so it is longer, more complex and more expensive. It is, however, very good optically, and can achieve very low 'size-of-source effect'.

4.4.1.4 Optical materials

For the design given above, scatter in the objective lens could be the major contributor to the instrument size-of-source effect uncertainty. The choice of objective material and any anti-reflection coating will influence the scatter. Special optical glasses are available and there is a range or properties for chemical stability, colour change, resistance to acid and alkali damage, the clarity of the glass and the internal scatter from bubbles and striations. Surface finish is assessed in terms of scratch-dig. Lower numbers are optically better. A typical standard grade would be 60:40, but better is 40:20 or even 20:10. In addition the deviation of the surface from the ideal is expressed as a fraction of wavelength: the optic is tested by observing interference fringes that give the difference in form between the optic and a test plate. $\lambda/2$ is rather poor, $\lambda/10$ is very good. There can be differences between data: if the surface form is specified as wavefront it is not the same as if specified on the surface – $\lambda/8$ wavefront is the same as $\lambda/4$ specified as the surface. Table 4.1 lists some typical lens materials with refractive index and absorption at specified wavelengths. In the visible range and up to $2\,\mu m$ there are a large number of different glasses with varying properties. The internal scatter will vary between glasses, which means different lens formulations from different suppliers will give better or worse size-of-source effect, however any improvement might be at the expense of changes in transmission from poor resistance to degradation from humidity or atmospheric pollution resulting in instrument drift.

Table 4.1: Common optical materials

Material	Wavelength range (µm)	Refractive index	Absorption per cm
calcium fluoride	0.13 – 10	1.40 (5 µm)	7.8×10^{-4} (2.7 µm)
sapphire	0.17 – 5.5	1.75 (1 µm)	0.3×10^{-3} (2.4 µm)
fused silica	0.18 – 2.2	1.45 (1 µm)	1×10^{-6} (1 µm)
silicon	1.2 – 15	3.42 (5 µm)	0.01 (5 µm)
germanium	1.8 – 23	4.00 (11 µm)	0.027 (10.6 µm)
BK7	0.35 – 2.5	1.51 (0.65 µm)	0.002 (0.65 µm)
zinc selenide	0.6 – 21	2.4 (10.6 µm)	5×10^{-4} (10.6 µm)

4.4.2   Reflecting optics

Mirrors have the advantage of being free of chromatic aberration. This makes them a good choice for broadband radiation thermometers. They are also relatively easy to manufacture to high standard without the problem of internal scatter contributing to the size-of-source effect. For long wavelength use they have very high reflectivity over a wide spectral range. The disadvantages are that it can be difficult to produce on-axis systems without adding plates or mounts that increase scatter (catadioptric design) or using off-axis components that increase bulk, cost and alignment complexity. Since there is no blocking of long wavelengths the possibility that heat could be concentrated onto particular components should be taken into account. Front surface mirrors are susceptible to degradation, particularly in areas where air quality might be poor, so need to be protected in some way.

4.4.2.1   Catadioptric

At its simplest a focusing primary mirror and a secondary mirror give a purely reflective system, with the secondary mirror mounted on a frame (the spider). A combined system can use a lens as part of the secondary mirror mount to give improved spherical aberration performance, but at the expense of reducing the long wavelength capability. The out-of-focus performance can cause unusual effects in size-of-source effect when there are hot sources not in the target plane. One potential advantage is that with opposed mirrors thermal expansion from changes in the thermometer temperature can be cancelled by suitable choice of material for the mount of the secondary mirror[2]. This can leave the optical path constant and unaffected by environmental changes. The folded light path can make for a very compact instrument.

4.4.2.2   Off-axis

Off-axis parabolic mirrors can provide a solution to the problem of mounting a secondary mirror, and introducing a filter and detector. A single mirror gives a system

---

[2]A similar effect can be achieved in lens-based systems, but not so simply.

free from chromatic aberration, and a pair can give very good optical performance if symmetrically arranged at 1:1 imaging, but can become very large and complex. In addition, off-axis parabolic mirrors are expensive.

4.4.3 Wavelength selection

The filter has other functions than just passing the required wavelengths; they limit the energy onto the detector and eliminate stray radiation from, for example, room lights.

- Interference filter. These have stacks of dielectric layers on a substrate that limit the range of transmitted or reflected light by constructive/destructive interference. They should be installed with the metallised surface toward the target. They are often used at an angle to reduce multiple reflections between optical components and hence ghost images. They are designed to be used at normal incidence so tilting will affect the filter wavelength. As they rely on the thickness of the dielectric layers any degradation will change the wavelength. Humidity can be a problem, and it is important to choose filters that are adequately sealed at the edges. Hard coated filters are also something to look for, where the layer density is higher using techniques such as ion-assisted deposition. For similar reasons it is usual to temperature stabilise the filter to prevent wavelength change due to thermal expansion. When exposed to high levels of radiance the filter can be heated leading to drift in the output. Additional heat rejection filters can be included ahead of the main wavelength-defining filter, or simple neutral filters can limit the overall radiance level.

  Interference filters can also be used to address specific issues, such as using a notch filter to remove sensitivity to humidity due to the water absorption band at 4.2 μm in the popular 3 μm to 5 μm band.

- Colour glass filter. Potentially these are less temperature sensitive than interference filters. While they can be combined as different glasses or in combination with the detector response to make a bandpass, they cannot produce narrow bandpass filters and do not have the range of wavelength and bandpass available with interference filters. They are, however, much cheaper. As they are based on optical glasses they similarly have a range of values for chemical stability, moisture resistance and optical quality.

- Grating monochromator. A ruled or holographic grating will reflect or transmit light at an angle that depends on the wavelength. Very narrow bandpass filters can be made using suitably sized slits to select the wavelength. The spacing or the rulings and the angle at which they are etched will influence the efficiency and wavelength range of use. Additional filters are likely to be required to remove higher-order wavelengths.

- Prism monochromator. A simple glass prism and a slit can select the wavelength.

In addition neutral density filters can be used to extend the upper range of an instrument. It is important that these are spectrally flat over the response of the instrument. It is also important that inter-reflections are considered and minimised.

4.4.4 Stray light reduction

The desired output signal of the pyrometer is that produced by the photodetector from radiation from the target area which has passed normally through the optical system, through the wavelength-selecting interference filter and possibly through neutral density filters if the source is at a high temperature. The total radiation entering the pyrometer covers a much wider wavelength range, and the quantum efficiency of the photodetector may be higher over part of this range than at the operating wavelength of the pyrometer. There are a number of techniques which may be employed in order to achieve the best level of performance. The primary optic should, as indicated above, be of excellent optical quality to minimise scatter. Lens design and construction have, of course, been thoroughly studied for other instruments, particularly for photography and astronomy, and good quality lenses are therefore available at a reasonable cost. For radiation thermometry, it should possess minimal spherical aberration and coma, and it should be achromatic around the operating wavelength or wavelengths. The optical system should of course be aligned both during construction and in use at the operating wavelength. If operation at more than one wavelength is envisaged, the lens should be achromatic at these wavelengths, i.e., the position of the circle of least confusion should not change. The interior of the instrument should be formed into independent compartments at each of the major stops or components, as a minimum at the field stop. The entrance stop in front of the primary optic should be extended to form a continuous shield at the front of the pyrometer so that only light from the vicinity of the target area can enter. Secondly, as stray light reaches the detector by means of reflections from the surfaces of the housing and of the optical components, it may be reduced by painting all internal surfaces with a matt black paint. Optical components with high reflectivity, such as thin film neutral density filters and interference filters, should be angled to remove the reflection from the main beam, and this is then absorbed on a black stop or surface. To do this effectively, the filters should be well separated, as otherwise the number of possible reflexions builds up rapidly. The tilting angle should not be large, around 5°, or the response of the thermometer will become sensitive to the polarisation of the incident radiation. The contribution to the signal from stray light may be estimated by placing a completely absorbing stop after the last optical element before the photodetector. This should not, however, affect the access of the stray light to the detector, which would reduce the apparent value, or increase it by allowing reflection from the back surface.

CHAPTER 5

Detectors

5.1 INTRODUCTION

In the previous chapter, we pointed out the importance of the radiation detector in the design and performance of radiation thermometers, and introduced the noise equivalent power (NEP) and the specific detectivity D^* as figures of merit for the comparison and selection of suitable detectors for a given application. In this chapter, we shall consider the characteristics of the different types of detectors in more detail, including not only these parameters and the responsivity, but the much wider range which may affect the choice for a particular instrument and which may influence the way in which the thermometer is used.

It should be emphasised that the values given in this chapter and in manufacturers' specifications are selected values from a range which may in some cases cover one or more decades, such is the variability in the characteristics of some detectors. Moreover, properties are measured for a limited set of what may be considered to be typical operating conditions, for example, at particular wavelengths or ambient temperatures. Estimates of their performance under other conditions introduces a further element of uncertainty. The figures supplied should therefore be treated as a guide; if more precise and reliable information is required, it is best obtained by direct measurements with the specific device types involved.

We may first comment that the detection process in thermometry is almost without exception the simplest possible, in that there is no mixing or heterodyning with other sources of radiation, or up-conversion to change the wavelength band into one which may be detected with greater ease or efficiency. The thermal radiation, which may be chopped in order to enjoy the benefits of a.c. techniques and to reduce the effects of $1/f$ noise which occurs in some detectors, is incident upon the active region of the detector, producing a voltage or a current as the output signal.

The other characteristics of detectors which may be important in the practical operation of radiation thermometers are:

(a) Linearity. If the output signal of the detector, with the unwanted contributions from the background radiation and the dark current subtracted, is directly proportional to the radiant flux incident on the active region, then the detector is said to be linear. This property is of major importance for the primary thermometers required to measure thermodynamic temperatures, or to realise the International Temperature Scale, where the absolute value for the ratio of the spectral radiances from two sources is required. For general purpose thermometers it is a desirable rather than an essential characteristic, as the relationship between the detected signal and the temperature of the target is strongly non-linear anyway and is a factor in the calibration. The degree of non-linearity may, however, vary with time or from one device to another, so that it presents the designer with another parameter whose effects must be studied and allowed for.

(b) Reproducibility. We include under this heading all unintended variations in the output signal which arise from changes in the characteristics of the detector or its immediate environment. In some cases, accurate corrections may be made. For example, the responsivity R_λ is often a function of the temperature of the detector. Errors from this source may be minimised either by stabilising the detector temperature, by measuring it and applying a correction to the output signal, or by incorporating a temperature-sensitive resistor network in the detector amplifier which varies the gain to compensate for the change in responsivity. The first of these solutions is to be preferred in principle, as other properties of the detector may also depend on temperature. The dark current, for example, increases exponentially with temperature, and, although it is usually subtracted from the signal at some point, the fluctuations remaining may well limit the resolution of the thermometer. However, in practice controlling the detector temperature adds significantly to the weight, power requirements and complexity of the instrument, and would be unacceptable for many portable and general purpose industrial thermometers. Monitoring the detector temperature allows corrections to be generated and applied separately for each of the temperature-dependent characteristics.

Similarly, the slow drift or ageing in the responsivity of many detectors may be allowed for by re-calibration of the thermometer at regular intervals. Where the instrument acts as a standard, the generation of a historical record of performance greatly adds to the confidence of the user in the validity of the current readings. If the drift is too great for this procedure to be satisfactory, an internal reference in the form of a small filament lamp or a black-body may be built into the device. The signal from the target is then compared directly to that from the reference source, so that variations in the detector characteristics are cancelled out. When the detector characteristics are unstable in the short term, or depend upon the history of illumination through hysteresis or fatigue effects, the technique is taken to its logical limit and the signals from the target and the reference balanced by switching quickly between them and reducing the resulting a.c. component of the output to zero. The effect of the resulting null detection system is to transfer the reproducibility requirements from the detector to the internal reference, which then effectively carries the calibration of the instrument. The improvements in modern radiation detectors have significantly reduced the proportion of radiation thermometers which are operated in this way.

(c) Response time. The speed of response of the detection system, which is usually but not always determined by that of the detector, is important in two situations: where the source temperature is varying rapidly and must be followed without undue loss of accuracy, and where the signal is being chopped, either as part of a null detection system as described in the previous paragraph, or to make use of a.c. detection techniques which avoid flicker noise and d.c. drifts. The spread of values found for common detectors is enormous, from several seconds for a large thermopile detector to less than 10^{-9} s for a fast semiconductor photodiode. Even within a given detector type, the response time will often depend upon the dimensions of the active area.

(d) Size and uniformity. Although the optical system of the radiation thermometer may be designed to match the target area and field of view onto the detector (or vice versa), the availability of appropriately sized photodetectors is of some importance. For example, some of the more exotic semiconductor devices are only available with active areas less than $1\,\mathrm{mm}^2$; it is necessary to focus the collected thermal radiation onto such a detector, with the result that small changes in alignment, possibly from mechanical vibration or stress, may produce significant variations in the output signal. A similar effect is produced by non-uniformities in the response across the active area of the detector. The errors may be reduced by 'flooding' the detector, that is, illuminating an area larger than the active area of the detector with a uniform or diffused beam, but at the cost of reducing the effective transmission or throughput of the instrument. Equally, very large detector areas are undesirable in that the unused regions contribute to the dark current but not to the detected signal. In addition, the time response of semiconductor and thermal detectors increases with the active area.

(e) Practical characteristics. The selection of a radiation detector suitable for a given application often involves the consideration of a number of additional practical points, for example, the size and weight of the detector and its associated mounting and, possibly, cooling system, amplifiers and power supplies. The ruggedness of the detector is important for instruments to be used outdoors and in heavy industry. And non-technical considerations, such as the cost and the commercial availability of the detection system, may turn out to be major factors.

5.2   CLASSIFICATION OF DETECTORS

It is conventional to divide detectors of thermal radiation into two groups, designated photon and thermal detectors. In the first, the absorption of a photon in the active region leads to the production of a charge carrier which may be collected by appropriate electrodes and hence recorded. In principle at least, these detectors can therefore detect single photons. Because the generation of the charge carrier requires a certain minimum of energy, photons with less than this are not detected. The spectral response curve therefore shows a threshold at a wavelength corresponding to this energy, falling to zero at longer wavelengths.

Thermal detectors, on the other hand, detect radiation by the heating effect upon absorption. A sensitive thermal detector must possess an electrical property, resistance or thermo-emf, for example, whose temperature coefficient is large. Although ideally thermal detectors could approach the Johnson noise limit in sensitivity, practical devices are much less sensitive. If the active area is blackened with a highly absorbing coating, it is possible to produce a device with a spectral responsivity flat over a very wide wavelength range, from the ultraviolet to the mid-infrared. The response time is limited by the thermal time constants within the device, and generally lies in the range from 1 ms up to 1 s.

We shall start by looking at the characteristics of the different types of photon detectors.

5.3 PHOTOEMISSIVE DEVICES

The emission of electrons by metals when irradiated by light, particularly at ultraviolet wavelengths, was discovered by Hertz during experiments aimed at validating Maxwell's equations. It was shown later by Lenard that the energy of the electrons emitted, measured from the voltage V required to stop them, was not dependent upon the intensity, as expected from the wave theories of light then generally accepted, but was related to its wavelength or frequency ν through an equation of the form

$$eV = h\nu - \Phi \qquad (5.1)$$

where Φ is the energy required to extract an electron from the interior of the metal. Einstein used the effect as evidence of the particle nature of light.

Photo-emitting materials can be semiconductors formed either from combinations of the alkali metals or of elements from groups III and IV of the periodic table, for example gallium arsenide. A simplified diagram of the band structure in such a material is given in Figure 5.1. An electron may be excited by the absorption of a photon from the valence band into the conduction band, leaving behind a hole. If it moves to the solid-vacuum interface and has sufficient energy to overcome the potential barrier E_A called the electron affinity of the material, it may be emitted into the vacuum carrying any excess as kinetic energy. There is therefore a threshold wavelength corresponding to the sum of E_A and the bandgap energy E_G beyond which the quantum efficiency falls to zero. The effective electron affinity may be reduced considerably, to around 0.5 eV, by coating the surface with a thin layer of a low work function material, usually either caesium or a caesium-oxygen mixture.

Within the photo-emitter, the electrons lose energy to the lattice as they diffuse

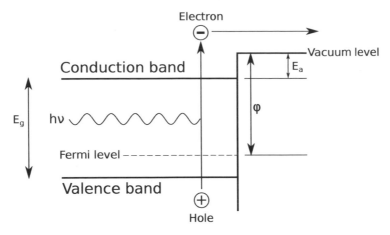

Figure 5.1: Band gap structure of the photoemission process.

towards the interface, eventually becoming 'thermalised', that is, having thermal energies. As a result, the diffusion length for 'hot' electrons, with sufficient energy to be emitted into the vacuum, is short, of the order of 10 nm. To obtain quantum efficiencies of any practical value, therefore, the absorption coefficient of the material must be such that electron excitation takes place within this distance of the emitting surface. This is not too difficult to achieve with opaque photo-emitters, where the light is incident on the emitting surface itself, as a high absorption coefficient alone is required. However, many practical photo-emitters are used in transmission, with the light entering from the opposite surface to that from which electron emission takes place. In this case, the properties of the photo-emitter, especially its thickness, must be carefully selected or monitored during manufacture so that a reasonable proportion of photons are absorbed close to the vacuum interface.

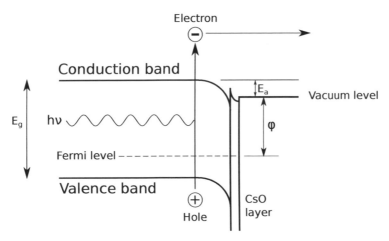

Figure 5.2: Negative electron affinity material band gap and surface potential.

A class of materials known as 'negative electron affinity' or NEA materials have been produced in which E_A has been reduced to zero and even negative values by controlling the band structure near the vacuum interface. The effect, shown diagrammatically in Figure 5.2, involves band-bending at the surface of strongly p-type semiconductors produced by coating with caesium or caesium-oxygen layers. As the vacuum level is below the conduction band, all electrons reaching the vacuum interface may be emitted, even if they are 'thermalised'. As a result, the diffusion length is much greater, of the order of a few micrometers, and quantum efficiencies may approach the theoretical maximum. Although the absorption coefficient tends to fall with increasing wavelength, the large diffusion length helps to keep the quantum efficiency high up to the threshold, which now corresponds with the bandgap energy E_G alone, since E_A is negative. In conventional photo-emitters the quantum efficiency tends to die away well before the threshold wavelength is reached.

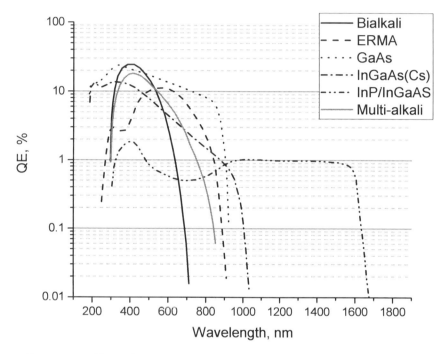

Figure 5.3: Typical spectral coverage of some photo-emitter detectors.

The quantum efficiencies of the commonest photo-emitters employed as photocathodes in vacuum photodiodes and photomultipliers are indicated as a function of wavelength in Figure 5.3. The various forms of the caesium antimonide Cs_3Sb photocathode have not been included, as they have generally been replaced by improved materials. The K_2CsSb bialkali photocathode, for use at wavelengths below the red, has the advantages of higher peak quantum efficiency and much lower thermionic noise level. The prismatic S-20 photocathode is an example of the technique of quantum efficiency enhancement. In many applications, sensitivity at the longer wavelengths is an advantage. As noted above, the fall in quantum efficiency towards the threshold is often a result of the reduction in absorption; it has been countered by increasing the effective path length of the light in the photocathode by introducing the beam at a large angle, particularly if total internal reflection then occurs. The prismatic S-20 cathode is laid down on a window which has been cross-machined to produce an array of small pyramids on the internal surface.

In general, a reduction in the bandgap and the electron affinity in order to obtain sensitivity in the red or near-infrared is accompanied by an increase in the level of thermionic emission, that is, the number of electrons emitted as a result of thermal rather than photon excitation. The level of thermal noise, N_t, from this source may be reduced by cooling the photo-emissive device, and follows the Richardson-Dushman equation

$$N_t = A\,T^2 \exp\left(\frac{-\Phi + bT}{kT}\right) \qquad (5.2)$$

where A is a constant equal to $7.5 \times 10^{24}\,\mathrm{m^{-2}\,K^{-2}\,s^{-1}}$, Φ is a thermionic work function and b its temperature coefficient. This representation appears to be more physically realistic than that which assumes that Φ is temperature independent and multiplies the equation by an arbitrary factor to represent the reflection of electrons at the surface.

The parameters Φ and b may be determined for a given photocathode by plotting $\ln(N_t/T^2)$ from the measured dark count or dark current as a function of $1/T$. At low temperatures, the thermionic emission dies away, leaving residual noise which varies much more slowly with temperature. This arises from a variety of sources, including cosmic rays, radioactive elements in the glass envelope etc. The residual noise level is usually sufficiently low that there is little point in cooling the device further; indeed, such action may be inadvisable in that it increases the likelihood of condensation at some point, or of damage from differential contraction of the glass and metal elements of the device. Indeed, the only photocathode which benefits from being cooled much below 0 °C is the S-1 type, which has a very high level of thermionic emission at room temperature and may need cooling to liquid nitrogen temperatures. However, tubes containing S-1 cathodes do not have a good reputation for reliability.

Other properties of the photocathode are changed on cooling, of course, particularly the quantum efficiency and spectral responsivity. Estimates of the typical temperature coefficients are given in Table 5.1. It is important to stress, however, that the temperature coefficient may vary widely with wavelength. For example, the extended red multi-alkali (ERMA III or S20R/S25) photocathode has a negative coefficient at short wavelengths, that is, the quantum efficiency increases as the cathode is cooled, while between 500 nm and 700 nm it is small. At long wavelengths, the threshold moves to shorter wavelengths as the temperature is reduced, with the result that above 820 nm the temperature coefficient becomes extremely large (5-10% per degree) and the device cannot be used for accurate measurements without very precise control of the photocathode temperature. The gallium arsenide opaque photocathode shows a similar effect at its threshold, around 900 nm. The quantum efficiency of the bialkali cathode also falls quite quickly with temperature, but the noise level from this cathode is sufficiently low at room temperature that

Table 5.1: Characteristics of common photocathodes

Cathode	Material	Maximum QE	Threshold (nm)*	T coeff. %/K	Workfunction Φ	eV/K
S-1	AgCs0	0.05	1200	−0.2		
Bi-alkali	K_2CsSb	0.30	660	−1.5	1.52	1.7×10^{-3}
S-20 multi-alkali	KNaCsSb	0.20	850	−0.3		
ERMA	KNaCsSb	0.20	900	−0.2		
	GaAs	0.25	930	−0.4	1.32	10^{-5}
	GaAsP	0.12-0.40	720-920			
	InP\InGaAsP	0.01	1400			
	InP\InGaAs	0.01	1700			

*Taken at a quantum efficiency of 0.0001.

little is gained in cooling. Moreover, its resistivity is high and increases with cooling, so that there is a danger of non-linear effects from changes in electrical potential across the cathode.

5.3.1 Vacuum photodiodes

A simple photodetector may be constructed from a photo-emissive material in an evacuated glass envelope, with a metal electrode at a potential of around 100 V acting as an anode to collect the emitted electrons. If the photocathode is opaque, it may be illuminated through the side wall of the envelope; if it is of the transmission type, it is usually formed on the inside of a plane window at the end of the device away from the base through which the electrical connections are made. The process of photo-emission is extremely rapid, of the order of 10^{-10} sec to 10^{-11} sec. The transit time of the photo-electrons between the cathode and the anode is also short, a few nanoseconds, so that the device is capable of very fast response, and is usually limited by the RC time constants of the circuits in the associated electronics.

The signal from the anode consists of the collected photo-electrons, while the dark current, from red-sensitive photocathodes, arises mainly from thermionic emission and from leakage currents, both internal and external, from the large potential difference between the anode and cathode. The former may be reduced by cooling, while the internal leakage currents may be controlled with guard electrodes. However, upon examination of (4.19) for the signal-to-noise power of the device, it is clear that the detector is amplifier noise limited for anything less than high light levels, and cooling is unnecessary and ineffectual. For these devices, the gain g, the excess noise factor Γ and the collection efficiency F may be taken as unity in value. If we assume a load resistance of 1 MΩ, an amplifier noise temperature of 300 K and a quantum efficiency of 0.02 at 800 nm, the noise equivalent power

$$NEP = \frac{h\nu}{\eta e}\left(\frac{4kT_N B}{R_L}\right)^{\frac{1}{2}} \tag{5.3}$$
$$= 1.1 \times 10^{-11}\,\mathrm{W\,Hz}^{-\frac{1}{2}}$$

for a bandwidth B of 1 Hz. This represents a very imperfect detector, and it is very rarely employed in accurate thermometry. For industrial applications, an additional disadvantage is the relative fragility of most vacuum photodiodes, and solid-state equivalents are universally preferred.

5.3.2 Photomultipliers

The limitations of the vacuum photodiode may be overcome by the addition of an electron multiplier within the glass envelope. This consists of a set of electrodes, called dynodes which increase the number of electrons by the process of secondary electron emission. A single electron with an energy of the order of 100 eV can release several electrons when it strikes a suitably prepared surface. The mechanism is not dissimilar to that of photoemission, except that the energy of the primary or incident electron may be sufficient to produce many electron-hole pairs. A fraction of the electrons, depending upon their initial energy, the depth of formation below the surface, and the electron affinity, will diffuse to the surface and be emitted. These are then accelerated in the inter-dynode potential field and directed towards the next dynode. The electrons cascade through the multiplier, increasing in number at each dynode by the stage gain, until they are collected by the anode.

There are three important characteristics of electron multiplication which make its use so valuable in photomultipliers. First, the process is almost noise-free, because the energy required to generate secondary electrons is much larger than the thermal energy of electrons in the dynodes. Secondly, the electron multiplier gain

$$g = \prod_{i=1}^{i=n} \delta_i \tag{5.4}$$

may easily be made as large as 10^6 for photomultipliers containing the usual number n of dynodes, between 9 and 13, as shown in Figure 5.4. The stage gain for typical secondary electron emitters reaches a maximum of 6 to 10 at an electron energy rather less than 1 keV. While the overall gain g may be increased by increasing the total potential drop across the multiplier, the maximum value is limited by ion and photon feedback rather than the maximum in the secondary electron gain. The onset of these feedback effects increases the noise level of the photomultiplier, and it should be operated at a lower gain.

The third important characteristic is the small transit time of the electron cascade through the multiplier. Typically, the delay between the emission of a photoelectron from the photocathode and the detection of the cascade at the anode is a

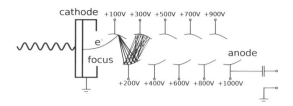

Figure 5.4: Sketch of linear showing electrodes and divider and cascade of electrons.

few tens of nanoseconds. The transit time spread, that is, the range of transit times of different electrons arriving at the anode, is even smaller, so that the electrons are detected as a pulse with a rapid rise-time of a few nanoseconds and a total duration three to four times longer. Together with the high gain, the effect is that the detection of one photon at the photocathode can give rise to a pulse at the anode which can readily be shaped and counted. This forms the basis of the technique of single photon counting.

In many applications, the rapid response time of the photomultiplier is not required, and the cascades of electrons arriving at the anode are averaged with an external RC network to give a smoothed anode current i_A. The current gain G of the photomultiplier, defined as the ratio of the anode and cathode currents when the tube is illuminated, is related to the electron multiplier gain g by

$$G = Fg \qquad (5.5)$$

where F is known as the collection efficiency, and describes the fraction of photoelectrons which fail to give detectable pulses at the anode. Losses may occur in the electron optical system between the cathode and the first dynode, or at the first dynode if no secondary electrons are produced or if they are not directed towards the second dynode. In d.c. operation, the signal from the anode is best taken through a current amplifier whose feedback resistor is switched to give the required sensitivity.

There are a number of different designs of electron multipliers in common use in photomultipliers, and examples are shown diagrammatically in Figure 5.5. Their characteristics are summarised in Table 5.2. The electric fields acting on the secondary electrons emitted from the dynodes of the venetian blind type are low and ill-defined at the dynode surface. As a result, the collection efficiency F is lower and the transit times greater for this type. At the other extreme, the potential fields and the shapes of the dynodes in the linear focused multiplier are designed to achieve efficient collection of the secondary electrons and to minimise the electron transit time. In some modern tubes, multiplication is provided by an internal channel multiplier. Although these can provide very high gains, the maximum photoelectron rate which they can cope with is limited by saturation along the individual channels, and they are therefore much less linear in response than the conventional designs shown.

Because the secondary multiplication process is statistical in nature, that is, a

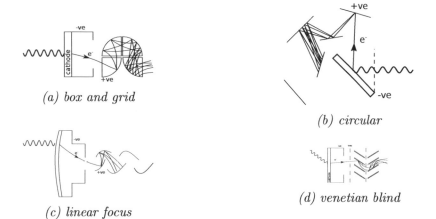

(a) box and grid

(b) circular

(c) linear focus

(d) venetian blind

Figure 5.5: Common dynode types.

Table 5.2: Pulse timing for common photomultiplier structures

Multiplier structure	Transit time (ns)	Rise time (ns)	Transit time spread (ns)
Box and grid	60–100	12–20	30–50
Venetian blind	60–100	5–10	15–20
Circular focused	30	1–2	2–5
Linear focused	30–60	2	2–6

primary electron with a given energy may produce only an integral number (including zero) of secondary electrons, the pulses collected at the anode are not all of the same amplitude or shape. It is often assumed that the multiplication process is described by Poisson statistics, that is, the probability of obtaining k secondary electrons from a dynode of gain δ is

$$P(k) = \exp^{-\delta} \frac{\delta_k}{k!} \tag{5.6}$$

However, pulse height distributions calculated with this assumption always show a smooth well-defined maximum, whereas experimental curves, with examples as given in Figure 5.6, may have no maximum or additional structure at high or low pulse amplitudes. A combination of different effects may be responsible for the wide variety of distributions found. For multipliers whose first dynodes have high secondary gain, as with the curve for a linear focusing multiplier in Figure 5.6, the presence of excess pulses of low amplitude appears to be related to the number of inelastically back-scattered primary electrons. Whereas a small number, of the order of 1%, of incident electrons are elastically back-scattered with no loss of energy, many more are back-scattered having lost a proportion of their original energy. The energy lost is converted into the equivalent number of true secondaries, which produce the smaller pulses. The fraction of back-scattered electrons is a function of the effective atomic weight of the surface, and varies for a gallium phosphide dynode from 0.24 to 0.32 depending on the degree of caesium coverage.

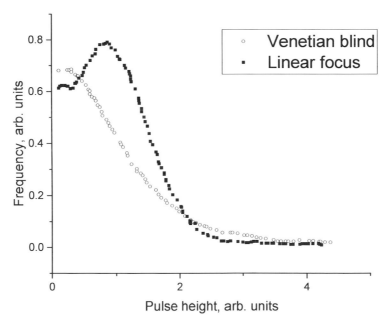

Figure 5.6: Typical pulse height distributions for venetian blind and linear focusing photomultiplier designs.

For venetian blind photomultipliers, which produce 'quasi-exponential' pulse height distributions, the non-Poissonian behaviour is probably a result of the weak electric fields around the dynode surface. As a result, some secondary and back-scattered electrons are not collected, and are lost from the multiplication process. It is significant that an increase in the inter-dynode potentials produces a noticeable increase in the collection efficiency F. On the other hand, electrons that are back-scattered in the right directions reach the next dynode with greater energy than the true secondaries, and may give rise to greater numbers of electrons at this stage.

The effect of the range of pulse heights is to increase the noise in the signal when the photomultiplier is operated as a current generator by a factor Γ. This is determined mainly by the gain δ_1 of the first dynode, and to a rapidly decreasing extent by the gain δ of any subsequent stages. In the case of Poisson statistics

$$\Gamma = \frac{1}{\delta_1} \frac{\delta}{\delta - 1} \tag{5.7}$$

For photon-counting detection, the anode pulses enter a discriminator which gives an output pulse of fixed height and duration when it is triggered. The value of Γ is then unity, as the amplitude distribution is single-valued. However, the counting efficiency – the proportion of anode pulses with amplitudes greater than the discriminator level – must be less than unity, and the collection efficiency F must be multiplied by this value to allow for the lost pulses. Although a peaked pulse height distribution is to be preferred for photon counting applications, counting efficiencies of greater than 90% can be achieved even with photomultipliers having quasi-exponential distributions.

Table 5.3: Typical dark current for various photocathodes.

Cathode	Anode dark current (A)	Cathode dark current (A)	D*	NEP (W Hz)
S1	1×10^{-6}	5×10^{-12}	5×10^{12}	6×10^{-13}
	4×10^{-6}	2×10^{-12}	1×10^{13}	3×10^{-13}
S4	1×10^{-9}	3×10^{-16}	6×10^{15}	2×10^{-16}
S10	3×10^{-8}	1×10^{-14}	8×10^{14}	6×10^{-15}
S11	7×10^{-10}	1×10^{-15}	1×10^{16}	3×10^{-16}
S20	1×10^{-9}	5×10^{-16}	8×10^{15}	2×10^{-16}
	1×10^{-8}	2×10^{14}	8×10^{14}	1.5×10^{-15}
Bialkali	2×10^{-10}	1×10^{-16}	4×10^{16}	6×10^{-17}
		3×10^{-16}	1×10^{16}	1×10^{-16}
GaAS	5×10^{-8}	5×10^{-14}	4×10^{14}	4×10^{-15}
	5×10^{-9}	2.5×10^{-15}	7×10^{14}	8×10^{-16}
InGaAs	3.5×10^{-8}	4×10^{-14}	6×10^{14}	2×10^{-15}
CdTe	5×10^{-10}	2×10^{-15}	8×10^{14}	1×10^{-15}
	1×10^{-10}	2×10^{-15}	4×10^{14}	3×10^{-15}

The overall effect of the large and noise-free gain provided by the electron multiplier is that the photomultiplier, unlike the vacuum photodiode, is not amplifier noise limited. At room temperature, tubes with red-sensitive photocathodes are dark current limited. Typical dark currents are given for different photocathodes in Table 5.3. These may be further reduced by cooling the photomultiplier, as discussed in the section on photocathodes above. The residual dark current, which is much less sensitive to the temperature of the photocathode, arises from cosmic rays, and natural radioactivity, principally potassium, in the glass envelope. Internal leakage currents, arising from the combination of the high voltages required for the electron multiplier and conducting materials such as caesium, which can coat the insulating supports for the dynodes, may also be significant in the current mode. With photon counting methods, the effects of leakage currents are usually negligible. Photomultipliers are extremely sensitive detectors at wavelengths for which the quantum efficiency is greater than about 0.01.

Unfortunately, the excellent performance of photomultipliers is not available without a number of accompanying problems. In the first place, they have a number of practical disadvantages. Because the internal structure is complex and not always very solidly supported, and is placed in an evacuated glass envelope, they are comparatively large, fragile and affected by mechanical vibrations. Ruggedised versions are available for some types at extra cost. They require in addition a high voltage power supply and a potential divider chain to establish the potentials on the various electrodes. Before being accelerated by the inter-electrode potentials, the electrons from photo- and secondary emission have low energy and are affected by stray magnetic fields. To avoid changes in sensitivity, the photomultiplier must be magnetically screened and operated away from strong magnetic fields. Altogether, photomultiplier detection systems are cumbersome, and more suited to laboratory instruments than those required for use in the field.

Secondly, because of the complexity of the device and the manufacturing process, the characteristics, even of a particular model from one manufacturer, vary widely. It is necessary, especially for critical applications, to select tubes from the available stock with the required characteristics. Most manufacturers offer this service at a moderate premium. While in an ideal situation the selected photomultiplier would have good red sensitivity or low noise, for example, and be acceptable in all other respects, it must be acknowledged that poor tubes exist. These must be eliminated by extensive testing of all relevant characteristics. The most frequent defect is the presence of excess noise generated by micro-discharges at sharp points on the electrode structure. This shows unpleasant noise statistics, similar to $1/f$ or even 'popcorn' noise, but may diminish and occasionally disappear with prolonged operation. Fortunately, some effects, such as ion after-pulses and cathode fluorescence, which detract from timing and pulsed experiments, are benign from the point of view of radiation thermometry, since the number of additional pulses varies linearly with the signal level.

In addition, a photomultiplier cannot be regarded as a stable system; it contains materials, principally caesium, whose distribution may vary with the current distribution through the device, and the potential fields may change as the surface charge or leakage current across insulators varies. In particular, the gain of the electron multiplier depends quite critically upon the surface composition of the dynodes, and changes produce an effect known as 'fatigue', which is present to a greater or lesser extent in almost all photomultipliers. It is most readily observed as a slow change, with a time constant of the order of several minutes, in the photocurrent after a sudden change in the level of illumination. The magnitude of the effect depends upon the photomultiplier type, the history of illumination and the anode current, and may be as large as several per-cent even for moderate operating conditions. While some types are as a class better than others in this respect, the level of fatigue under given conditions will again change from tube to tube.

Because the photon count rate is relatively insensitive to changes in the gain of the multiplier, the error from fatigue is considerably reduced from that observed with d.c. techniques. By measuring both the anode current and the output pulse count rate from the photomultiplier, a further correction may be derived which reduces the error to very low values. In commercial applications, the level of fatigue may be such that null techniques must be employed, that is, the radiance from a reference source is adjusted to match that from the target so that the level of illumination of the photocathode does not change when the thermometer is switched from one to the other. Under these conditions fatigue and non-linearity in the detector are negligible.

There are a number of general rules and suggestions which may be of assistance to those assembling and operating a photomultiplier system for the first time. As it is recommended that the gain and hence the maximum anode current should be restricted to quite moderate values, a resistive potential divider with a total resistance of the order of $1\,M\Omega$ should be adequate. This gives a divider chain current of around $1\,mA$ or less, which means that the dynode potential distribution is not significantly changed by the secondary emission currents, which of course increase at each stage towards the anode. It is good practice to increase the cathode-first dynode voltage compared to the remaining stages, especially in photon-counting systems, as this increases the collection efficiency and the first dynode gain. With a reasonably well-stabilised power supply, it is not necessary to further stabilise the voltages on the first few stages, unless the tube is to be run at very low gains. Although it is often recommended that the cathode should be operated at earth potential, and the anode at (positive) high voltage, it is practically more convenient to connect the anode to earth, especially for d.c. output signals. To avoid spurious signals and possible damage to the photocathode, the mu-metal screen should be connected to cathode potential (through a high resistance for the safety of the user), and metal components at earth potential should not be allowed in the vicinity of the cathode.

For photon counting, the last three dynodes should be de-coupled with high voltage disc ceramic capacitors, as shown in Figure 5.7. For very fast photomultipliers, it is sometimes necessary to include damping resistors in the leads to suppress ringing after the output pulse. It should be stressed that the rise time for many photomultipliers is only a few nanoseconds, and the length and layout of the wiring of the last few stages should reflect this. If the anode load is $50\,\Omega$, a short coaxial lead may be used to connect to the pre-amplifier and discriminator, which should be mounted as closely as possible to the anode load. It is not necessary to make the dead time of the discriminator very short, less than $10\,\text{ns}$ for example, as a correction to the count rate may be made if required. The dead time may be chosen to eliminate the chance of double pulsing from any residual ringing on the output pulse when low discriminator levels are being used.

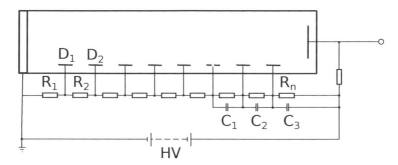

Figure 5.7: Dynode divider chain and bypass capacitors.

Photomultiplier bases, if used, should be kept clean and free from solder flux or grease from the fingers to minimise external leakage currents which may be significant in d.c. operation. For the same reason, cooled tubes should have the region around the base connections sealed, with silicone rubber for example, to prevent traces of condensation accumulating there. Commercial housings usually have a good sliding fit between the photomultiplier assembly and the container as a first line of defence, and should incorporate radio frequency screening at the junction in order to prevent triggering of the discriminator from broadcast interference.

In general, photomultipliers should be run at a current gain of between 10^5 and 10^6. Most are supplied with a check card which indicates the supply voltage necessary to achieve this gain; alternatively, the published specifications may provide a guide. When a photomultiplier is switched on for the first time, or after having been exposed to ambient light levels, the supply voltage should be increased slowly from well below this value while the output dark current or count rate is monitored. In some cases, the initial dark current is very high, and should be allowed to decrease to a reasonable level before the voltage is increased towards the required value. The initial stabilisation period may take anywhere from 10 minutes to 24 hours, and no attempt should be made to cool the tube until stable readings have been obtained. The final dark current or count rate should be compatible with the figures supplied by the manufacturer. If it is too high, there may be some leakage of light onto the photocathode, which may be shown by covering the housing with a black cloth or switching off the room lights.

For photon-counting systems, the next step is to check the discriminator setting and pulse height distribution with weak illumination of the photocathode. As the discriminator setting is increased from its minimum value, the signal count rate (with the dark count subtracted) should trace out the integral of the pulse height distribution. The zero level may be found by detecting the setting at which the discriminator is triggered by noise with the photomultiplier off. If this point is unavailable, it must be assumed that the zero on the discriminator level potentiometer has been set correctly. The aim is to set the discriminator at the minimum level which gives no counts with the photomultiplier off, and transmits 90-95 % of the signal pulses when it is on. It may be necessary to insert a suitable amplifier if the discriminator is insensitive; it is usually undesirable as an alternative to increase the gain of the photomultiplier above about 10^6. Some commercial discriminators have a second level which can be used to eliminate the large pulses from cosmic rays and after-pulses. In general, these are of little benefit in thermometry applications, and may complicate the measurement of the correction for the dead time of the counting system. One further test that may be made is to increase the supply voltage by 10-20 %. The count rate should increase slightly, because a larger fraction of the pulse height distribution triggers the discriminator. If the increase is much larger, it is a sign that ion or photon feedback is increasing the noise level, and the supply voltage should be reduced to a lower value.

5.4   PHOTOCONDUCTIVE DEVICES

A photoconductive detector is a relatively simple device, consisting of a sample of a uniform semiconductor material with two electrodes attached. The absorption of a photon of sufficient energy generates an electron-hole pair, which changes the effective conductance or resistance of the detector. A bias voltage is applied to collect the charge carriers, and the signal measured across a load resistor R_L as shown in Figure 5.8. There are two rather different classes, illustrated in Figure 5.9: intrinsic, where the photoconductivity is an inherent property of the semiconductor, and extrinsic, where it is developed by doping the material with selected impurity atoms. In extrinsic photoconductors, p-type (acceptor) impurity levels generate holes as charge carriers, while n-type (donor) levels produce electrons. As with photoemissive detectors, there is a long wavelength cutoff given by the ionisation energy of the impurity level, or by the bandgap in the intrinsic case.

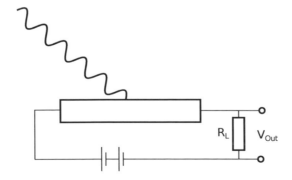

Figure 5.8: Basic electrical circuit for photodetectors.

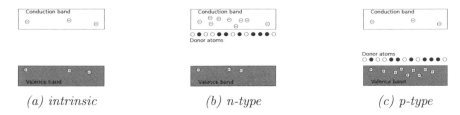

(a) intrinsic   *(b) n-type*   *(c) p-type*

Figure 5.9: Intrinsic and extrinsic semiconductors.

The charge carriers move in the semiconductor with a velocity v given by the product of the electric field E and the carrier mobility μ. Because the carrier lifetime τ may be much longer than the time required for the carrier to travel the distance l between the electrodes, photoconductors show a gain g given by

$$g = \frac{E\mu\tau}{l} \tag{5.8}$$

In effect, because electrical neutrality must be maintained when the carrier is collected by the appropriate electrode, it is re-introduced at the oppositely charged electrode and passes through the device several times. For intrinsic photoconductors, both carriers are present, so that

$$g = \frac{(\mu_n\tau_n + \mu_p\tau_p)E}{l} \tag{5.9}$$

although in general one term dominates the expression, so that the overall result does not differ greatly from the extrinsic case.

The absorption coefficients of most intrinsic photoconductors is high, above $10^6\,\mathrm{m}^{-1}$ so that a thin layer a few micrometres thick is sufficient to give peak quantum efficiencies of greater than 0.8. The device is formed upon a suitable semi-insulating substrate for mechanical strength, which is selected to match the atomic spacing of the photoconductor in order to prevent dislocations and similar effects. In some types, the quantum efficiency is optimised by illuminating the photoconductive layer through a transparent substrate, which may be anti-reflection coated for the wavelength range required. As a result of the low density of impurity atoms in an extrinsic detector, the absorption coefficient is much lower, around $10^3\,\mathrm{m}^{-1}$ to $10^4\,\mathrm{m}^{-1}$, and device thicknesses of about 1 mm are required. The signal current from an incident monochromatic power P_S is, including the gain term g

$$i = \frac{\eta e g}{h\nu}P_S \tag{5.10}$$

Changes in the temperature of a photoconductor may have a variety of effects. For extrinsic materials with low ionisation energy for the impurity atoms and hence with response at long wavelengths, the sensitivity at a given wavelength decreases as the temperature increases. This is partly a result of a greater proportion of the impurity atoms being thermally ionised, and it is possible for the sensitivity to fall to zero at quite moderate temperatures. In general, such photoconductors must be operated at very low (e.g. liquid helium) temperatures, and this restricts their application to laboratory measurements. In intrinsic photoconductors, the increase in the thermal noise generated within the device may have a pronounced effect upon the sensitivity. For example, D^* for an indium arsenide detector at 6.5 μm increases from 3×10^8 at ambient temperatures to 3×10^9 at −40 °C. Other effects may alter the shape of the spectral response curve; cooling reduces the separation of the valence and conduction bands, and hence the cutoff wavelength. In addition, the carrier mobility is reduced, affecting both the gain and the time response.

Because solid-state detectors may be of low thermal capacity, it is often possible and convenient to cool them with Peltier thermoelectric devices, which have the advantages of small size, good temperature stability and convenience of operation. If steps are taken to minimise the thermal load, the power consumption is reasonable, and multi-stage coolers may reach −80 °C.

In discussing the noise characteristics of these detectors, we note first that the carriers are in thermal equilibrium with the lattice, as the time between collisions is much shorter than the carrier lifetime. The thermal or Johnson noise current is therefore inversely proportional to the dark resistance of the conductive layer. The current noise, as discussed in Chapter 4, is given by

$$i_n^2 = 2HgeiB \qquad (5.11)$$

where g is the photoconductive gain and H the noise enhancement factor. The latter arises from the statistical processes involved in the production and removal of the charge carriers in the photoconductor, and is known as generation-recombination (or g-r) noise. The value of H may be shown to be equal to 2 if the lifetime or time constant for one particular process dominates, that is, that other generation or recombination effects have much shorter time constants. Note that the current i is the total current flowing within the device. We note that the load resistance should in this case be replaced by the parallel combination of R_L and the detector resistance. Substituting typical values into the equations for the signal-to-noise power ratio, we find that photoconductive detectors are generally dark current-limited.

There are many other sources of noise in photoconductive detectors, generated for example at the surface or at electrical contacts. They are usually negligible except at low frequencies where they may have a $1/f$ dependence. For optimum detectivities, these devices are operated at frequencies above 100 Hz and often around 1 kHz. The entries in Table 5.4 indicate the frequency at which the associated values are determined. Table 5.5 and Table 5.6 gives the spectral coverage for some available photoconductive materials.

5.5 SEMICONDUCTOR PHOTODIODES

A device containing a junction between n- and p-doped semiconductor materials acts as a rectifying diode. Under normal conditions, the majority of the doping atoms in such materials are ionised, with the result that there are free holes in

Table 5.4: Properties of selected photoconductive detectors

Material	Size (mm^2)	Temp (^{o}C)	Responsivity (V W^{-1})	Detectivity	Freq. (Hz)	Dark resistance (MΩ)	Response time (μs)
PbS	5 × 5	25		2 × 10^{10}	100	0.3–20	100–400
	1 × 3	−10		2 × 10^{10}	100	0.3–20	800
	10 × 10	25		9 × 10^{10}	750	1–10	40–500
	5 × 5	−30	1–6 × 10^5	2 × 10^{11}	600	1.5–2.5	1000
	10 × 10	−80	5 × 10^6	6 × 10^{11}	780	0.5–5	2000–5000
PbSe	1 × 5	25		1 × 10^9	100	0.1–2	2
	4 × 4	25	1–6 × 10^3	2.5 × 10^9	750	0.5	1
	5 × 5	−30	5–13 × 10^3	1 × 10^{10}	1000	1–3	12
	10 × 10	−80	5 × 10^5	3 × 10^{10}	780	10–50	10–40
InSb	2 × 2	−20	20	6 × 10^8	1000	4	0.2
HgCdTe	2 × 2	−80	3 × 10^4	3 × 10^{10}	10000	10^{-3}	0.2–10

Responsivity and detectivity measured at peak wavelength.

Table 5.5: Intrinsic semiconductor devices

Material	Device	Energy gap (eV)	Peak wavelength (μm)
HgCdTe	c		2–30
PbSnTe			2–15
InSb	c	0.23	5.4
PbSe		0.27	4.6
PbTe		0.3	4.1
InAs		0.33	3.8
Te	c	0.33	3.7
PbS		0.34–0.37	3.5
ZnSb	j	0.56	2.2
SiGe		0.67	1.9
GaSb	j	0.78	1.6
Si		1.14	1.1
InP	j	1.25	1
GaAS		1.4	0.89
CdTe		1.45	0.86
AlSb	c	1.6–1.7	0.75
CdSe		1.74	0.71
GaP		2.25	0.55

c = photoconductor, j = junction photodiode.

Table 5.6: Extrinsic photoconductors

Material	Cutoff wavelength (μm)	Material	Cutoff wavelength (μm)
GeAu	7	SiIn	7
GeHg	11	SiGe	17
GeCd	21	SiBi	17
GeCu	23	SiAl	18
GeZn	40	SiAs	23
GeBe	40	SiP	28
GeB	110	SiB	28
GeGa	120	SiSb	29
GeLi	140		

the valence band of the p-type semiconductor, and free electrons in the conduction band of the n-type material. In both, the Fermi levels are close to the energy levels of the dopant atoms, as indicated in Figure 5.10. When the junction is formed, the holes from the p-type material and the electrons from the n-type diffuse towards the junction and combine, producing a depletion region in which there are few mobile charge carriers, that is, a layer of high resistance. The movement of charge produces a potential difference across the junction which is maintained by the high resistance of the depletion layer. The junction behaves as a small capacitor, with two charged regions separated by a high resistance layer, whose capacitance depends upon the area of the diode and the thickness of the depletion layer. With no external voltage applied across the diode, through the metal contacts shown in Figure 5.10, it may be shown that the condition for equilibrium is that the Fermi level is constant through the device, as shown in the diagram.

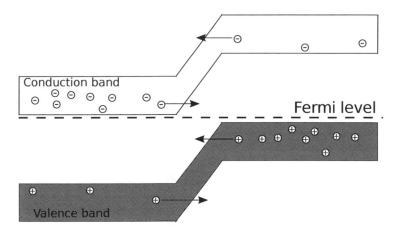

Figure 5.10: p-n junction band structure in photodiodes.

The application of a voltage across the diode changes the extent of the band bending obtained. If the polarity of the voltage is such that the bending is reduced, so that current may flow more easily, the junction is said to be 'forward biased', while if the potential barrier is increased, the diode is 'reverse biased'. The current components which flow through the junction are of two types; a diffusion current is produced by those electrons in the conduction band of the n-type material which have sufficient energy to overcome the potential barrier. This will clearly depend on the applied voltage V,

$$i_{nd} = i_{nd}^0 \exp\left(\frac{eV}{kT}\right) \tag{5.12}$$

where the constant term i_{nd}^0 represents the electron diffusion current with zero bias. A generation current i_{ng} arises from the thermal excitation of electrons from the valence band to the conduction band in the p-type material. As these carriers are assisted across the junction by the potential gradient, the generation current is independent of V. Two similar contributions i_{pd} and i_{pg} are due to the movement of holes across the junction, with the same dependence upon the applied voltage. When V is zero, the total current must be zero, that is,

$$i_{nd}^0 + i_{ng}^0 - (i_{pd} + i_{pg}) = 0 \tag{5.13}$$

so that at any other voltage V the total current i through the junction is given by

$$i = i_s[\exp(eV/kT) - 1] \tag{5.14}$$

where i_s is known as the saturation current and is given by

$$\begin{aligned} i_s &= i_{nd}^0 + i_{pd}^0 \\ &= i_{ng} + i_{pg} \end{aligned} \tag{5.15}$$

The saturation current is typically in the range $10^{-7}\,\mathrm{A}$ to $10^{-9}\,\mathrm{A}$ for silicon diodes.

The absorption of photons of sufficient energy in or around the depletion region generates hole-electron pairs which are separated by the potential gradient, to produce a signal current which in terms of the conventions adopted here is negative, i.e.,

$$i = \frac{\eta e P_S}{h\nu} + i_s \left\{ \exp\left(\frac{eV}{kT}\right) - 1 \right\} \tag{5.16}$$

The effect upon the electrical characteristics of the photodiode is shown in Figure 5.11. In the absence of thermal radiation, the variation of the total current i with bias voltage V is given by the curve labelled '$P_S = 0$'. The additional photocurrent produces a family of similar curves displaced downwards with increasing incident power.

Depending upon the detector characteristics required, the electrical circuit attached to the diode will vary. If it is to be used in the photoconductive mode, reverse

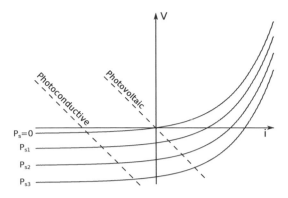

Figure 5.11: Electric characteristics of photodiode.

bias is applied, and the circuit is similar to that shown in Figure 5.8. While the bias decreases the capacitance of the junction and the transit time of the carriers, improving the response time of the device, an additional leakage current is generated which may reduce the detectivity. Photodiodes, particularly silicon types, are more usually connected in a photovoltaic mode with no applied bias. If the resistive load R_L is too large and the voltage developed across it becomes significant, then the signal, represented by the load line with slope $R_L = V/i$ in Figure 5.11, depends non-linearly upon the incident power level. Indeed, as the output voltage approaches the forward conduction voltage of the diode, the signal saturates. To ensure that optimum linearity is achieved, the photodiode should work into a short circuit. A current amplifier with a low input impedance is normally employed to achieve this situation while producing a useable output signal. Hybrid detectors are available in which the amplifier and the detector are contained in the same package.

In general, the response time of a simple photodiode will depend on a number of factors, principally the RC time constant of the external circuit, which will include the junction and stray capacitances, the transit time of carriers across the depletion layer, and the time taken for carriers formed away from the depletion layer to diffuse into the region of high potential gradient. While the transit time is usually small, around 1 ns to 10 ns, the diffusion time may be a thousand times longer. Depending upon the relative importance of the two effects, a given device may therefore show two – and sometimes more – time constants.

Improved performance may be obtained by modifying the design of the photodiode to include a layer of intrinsic material between the highly doped regions. Typical device structures, known as PIN photodiodes, are shown in Figure 5.12. The thickness of the intrinsic layer is selected to give high absorption at wavelengths in the response band. At long wavelengths, this is limited by the bandgap of the material, while radiation at short wavelengths is absorbed near the surface of the device, where the processes of carrier recombination are rapid and the responsivity therefore small.

The inclusion of the intrinsic layer reduces the effect of the diffusion component upon the response time, giving a faster device; it also reduces the leakage current

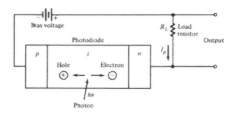

Figure 5.12: Structure of PIN photodiode.

and hence the noise, and improves the linearity in the photovoltaic mode. The best silicon photodiodes of this type show a remarkable range of good characteristics, including

- Long-term stability, so that the response changes typically by less than 1 % over a period of five years.

- A response band extending from 300 nm up to 1100 nm, with a quantum efficiency at the maximum which approaches 100 %.

- A response uniform over most of the sensitive area of the device to within 1 %. In many cases, the limiting effect may be from interference in the window of the package or can.

- The signal output is linear up to a current of about 1 µA, enabling incident powers over a very wide range, for example from 10^{-12} W to 10^{-2} W to be measured accurately and without difficulty.

Heterojunction photodiodes consist of semiconductors of different bandgaps. An example consists of InGaAsP epitaxially grown upon an InP substrate of high surface quality to reduce surface absorption and recombination effects. It is necessary to match the lattice constants of the semiconductor and substrate in order to reduce leakage currents. By varying the composition of the quaternary semiconductor, it is possible to obtain bandgap energies in the range from 0.73 eV to 1.35 eV, that is, with a maximum long wavelength threshold of 1.7 µm. Surface losses may be reduced by illuminating the junction through the substrate layer, which is transparent above 0.95 µm. The absorption coefficient of the semiconductor, around 10^6 m^{-1}, requires an active region thickness of about 2 µm to achieve high quantum efficiencies.

The total shot noise contains contributions from the signal and background photocurrents, from the dark or saturation current i_s and from the forward current induced by the bias voltage V. These are independent and must be added together, irrespective of their direction, that is,

$$i_N^2 = 2e \left[\frac{\eta e (P_S + P_B)}{h\nu} + i_s + i_s \exp \left(\frac{eV_A}{kT} \right) \right] \qquad (5.17)$$

When the diode is unbiased, the equation for the NEP is very similar to that for a vacuum photodiode (5.3).

5.6 AVALANCHE PHOTODIODES

When the reverse bias on specially designed photodiodes is increased, carriers generated thermally or from the absorption of photons can gain sufficient energy to produce additional hole-electron pairs in the depletion layer by ionisation of the lattice atoms. These carriers can then continue the process, giving internal current gain characterised by a multiplication factor M. The overall effect is similar to that within a photomultiplier with a dynode gain of 2, except that the effective number of dynodes increases with the bias voltage. It is often known as avalanche multiplication, and hence the devices are called 'avalanche photodiodes' or APDs.

At high voltages, it is possible for the current to become very large and avalanche breakdown occurs. As the device characteristics are then very non-linear and unstable, this situation must be avoided. The presence of dislocations or inhomogeneities in the material of the multiplication region can lead to local breakdown, and this limits the maximum overall gain which can be obtained to around 100.

The multiplication process is characterised by the ionisation coefficients α_n and α_p for the electron and the hole, respectively, and by the ratio k of the smaller of these to the larger. Being statistical in nature, the process introduces an excess noise factor Γ given approximately by the equation

$$\Gamma = kM \left(2 - \frac{1}{M} \right) (1 - M) \tag{5.18}$$

When the M is reasonably large, two particular cases are often employed to illustrate the behaviour of Γ. If multiplication takes place primarily through one of the two carriers, that is, k is small, then Γ is approximately equal to 2, its minimum value. On the other hand, if both carriers are equally effective (that is, $k \approx 1$) then Γ is close to M, and the noise introduced is large. To produce effective avalanche photodiodes, therefore, it is necessary to find materials or manufacturing techniques which yield low values of k.

The expressions for the signal and mean square noise current are similar to those for a photomultiplier:

$$i_s = \frac{e\eta}{h\nu} M P_S$$

$$i_N^2 = \frac{\left(\frac{e\eta}{h\nu} P_S \right)^2}{2e \left[\frac{e\eta}{h\nu} (P_S + P_B) + i_D \right] BM^2 \Gamma + \frac{4k_B TB}{R_L}} \tag{5.19}$$

As most avalanche photodiodes operate at wavelengths in the visible or near infrared, they are not usually background limited. At moderate values of M, the amplifier noise becomes negligible, so that these devices are normally dark current limited.

Avalanche photodiodes may be fabricated from silicon, germanium and III-V semiconductors. The spectral response curves and the peak quantum efficiencies are similar to those of the corresponding more conventional photodiodes.

5.7 THERMAL DETECTORS

In the case of thermal detectors the detector is heated by radiance from the source, and the temperature of the detector itself is measured. As there is no threshold energy to be reached by the photons arriving from the target, thermal detectors can respond to long wavelength radiation. An ideal thermal detector is limited by shot noise from the detection process and from fluctuations in the radiation field,

$$i_N^2 = 2eB\frac{\eta e P}{h\nu} + \left(\frac{\eta e}{h\nu}\right)^2 \overline{\Delta P^2}$$
$$= 2eB\frac{\eta e P}{h\nu}(1 + \eta e f) \tag{5.20}$$

where f is the occupancy factor. An ideal thermal detector approaches the sensitivity of a photon counter but thermal detectors are generally slow.

5.7.1 Thermopile

Thermopiles are a class of detectors that rely on the Seebeck effect to convert radiation to equivalent electrical energy. In the Seebeck effect a temperature gradient along a conductor leads to a potential difference (emf) between the hot and cold ends, the magnitude of which is material dependent. By pairing two dissimilar conductors which are subjected to the same temperature gradient and are connected together at one end, an emf can be generated that depends on the temperature difference between the hot and cold ends of the conductors. Thermopiles use an array of thermocouples, usually connected in series, with the hot junctions on one side of the detector and the cold junctions on the other. They therefore act as a temperature difference sensor. The hot side incorporates an absorbing surface that warms on being exposed to thermal radiation. The surface should have high emissivity, and absorption can be increased by using a hemispherical mirror to reduce reflected or re-radiated heat loss, or by operating in a vacuum. Once such a detector has thermalised, its output will be zero, so the gradient needs to be maintained. This can be done by minimising the heat flow by using thin wires, by having a heat sink to remove the heat from the cold junctions, by actively controlling the cold junction temperature or by chopping the radiance. Thermopiles have a spectral response that depends on the blackness of the thermal coating and the transmittance of any window used.

5.7.2 Bolometer

A bolometer measures the temperature rise of a sensing material by measuring the change in its electrical resistance, i.e., the voltage V produced by a current I. The characterising parameter is $b = 1/V(dV/dT)$. If the sensing element has a thermal link with conductance G to the surroundings at T_0 then the noise voltage is given by

$$V_N^2 = 4k_B T_N RB + \frac{4I^2 b^2 R^2 k_B T_0^2}{G}\left(1 + \frac{I^2 R}{GT_0}\right) \tag{5.21}$$

From the signal

$$V_S = IbRP_S G^2 \tag{5.22}$$

the NEP and S/N can be calculated. Microbolometers are used as array detectors in thermal imaging cameras. Where thermopiles are based on thermocouples, bolometers are based on resistance thermometers.

5.7.3 Pyroelectric detector

A pyroelectric sensor produces a voltage when the temperature of the sensor changes. The atomic lattice structure is stressed by the change in temperature and generates a voltage in a similar manner to the piezoelectric effect. Pyroelectric detectors have high sensitivity and good response times. The signal

$$\begin{aligned} i_S &= AK_p \frac{dT}{dt} \\ &= AK_p \omega P_s/G \end{aligned} \tag{5.23}$$

is proportional to the rate of change of temperature so that it has to be chopped, and the mean square noise current is given by

$$i_N^2 = \frac{4k_B T_N B}{R} + 4K_p^2 A^2 \omega^2 \frac{k_B T}{G} B \tag{5.24}$$

which allows the signal-to-noise to be determined.

5.8 DETECTOR LINEARITY

One of the most important factors in a detector is its linearity. This will decide how an instrument can be used and how it can be calibrated. For instance, in order to measure a temperature absolutely with a radiation thermometer, that is, without reference to a calibrated instrument or device, there are two essential requirements. The first is the need for one or reference sources with known or defined temperature. The second is the ability to measure absolutely ratios of spectral radiance at the operating wavelength of the thermometer. Given these, and assuming of course that the spectral characteristics of the thermometer components are known or may be measured, the unknown temperature of a source may be obtained directly from its spectral radiance ratio with the reference source. The major difference between primary and other radiation thermometers is the requirement for accurate determination of radiance ratios. The approach taken to achieve this depends critically upon the behaviour of the photo-detection system. At a given wavelength, all other components within the pyrometer will transmit a fixed proportion of the incident radiation, and do not therefore affect the spectral radiance ratio from two sources at different temperatures (due allowance must of course be made for the differing spectral distributions of the radiation from the sources over the finite bandwidth of the thermometer. The output signal I from the photo-detector, including in this any subsequent stages of amplification, may in general however be written as a polynomial in the incident intensity,

$$I = a_0 + a_1 S + a_2 S^2 + \ldots \tag{5.25}$$

If the contributions from several terms are significant in terms of the accuracy required, then each coefficient a_i must be determined in order to calculate the radiance and hence the radiance ratio. This procedure is made very difficult in practice because they tend not to be stable, but to depend upon several factors, including the temperature of the detector, the applied voltage and even the past history of illumination. If this is the case, it is much simpler to operate the system as a null detector, that is, to attenuate the radiation from the source at higher temperature until the measured signal is equal to that from the lower temperature source. The radiance ratio is then given by the attenuation of the device used, which to be of use must be measured or calculated in some other way.

In one important case there is an even simpler method. If the detector is linear, the output signal is given by

$$I = a_0 + a_1 S \tag{5.26}$$

We note first that a_0 is the background signal, i.e., that detected when no radiation from the source reaches the photo-detector. Although it may well vary from one set of measurements to the next, it may easily be obtained by closing a shutter in front of the detector and subtracted from signal measurements made around the same time. If we assume that the signals have been corrected in this way, the radiance ratio is now given directly by

$$\frac{S_2}{S_1} = \frac{I_2}{I_1} \tag{5.27}$$

and the coefficient a_1, assuming that it remains stable over the time taken for this set of measurements, cancels out. Therefore, if it can be demonstrated that the detector behaves linearly within a useable range of experimental conditions, then the derivation of the radiance ratio from the measured signals is particularly simple. It should be emphasised that the constant of proportionality a_1 need only be stable over the period required to determine each ratio I_2/I_1, and it may drift or change between measurements without affecting the validity of (5.27).

Some of these advantages may be retained if the departures from linearity are not too great, as corrections for this may be applied. A rule of thumb is that the correction should not exceed ten times the accuracy required. The correction may be expressed in graphical or tabular form as a function of the signal level and the operating conditions of the detector. For analytical purposes, it is often convenient to include the quadratic term from (5.25), and to express the signal (with the background subtracted) as

$$I = a_1 S + a_2 S^2 \tag{5.28}$$

where the second term is always much smaller than the first.

The two approaches described are in general complementary. Devices which give a known or calculable attenuation may be employed either in a null measurement technique or to check the linearity of the detection system. A subsidiary use is to change the operating range of the pyrometer, at the same time ensuring that the thermal effects of the radiation entering the pyrometer are kept within bounds, and that the detector remains within its specified operating range.

Devices which enable absolute radiance ratios to be established fall into two groups. The first includes those, such as sectored discs, neutral density filters and crossed polarisers, which reduce the radiance in a given beam by a fixed amount. In the second group, separate beams from one or more sources are combined in such a way that the signals from a single beam or any combination may be measured. If the detector is linear, then the total signal observed with a combination of beams should be equal to the sum of the signals obtained from separate single beams. It should be emphasised that each technique has its errors and limitations, and these should be assessed with care. If possible, the results from two distinct methods should be compared to give confidence in the validity of the results.

5.8.1   Sectored discs

These consist of a disc with arcs or slots cut into it. When the disc rotates, the beam from the source, which is usually focused at the plane of the disc, illuminates a circle on the disc about its centre of rotation. The transmission of the disc is given by the ratio of the length of the slot or slots along this circle to the total circumference, assuming that the disc rotates at a uniform angular velocity. Alternatively, it is given by the fraction of the time that the beam can pass through the disc. The advantages of these devices are that the transmission of the disc may be independently measured from its geometry, and that this should be independent of wavelength, as the dimensions are large enough that diffraction effects are negligible. Also, being usually fairly massive and made of metal, they are unaffected by the heating effects of the radiation from the source. If the ends of the slots are placed radially, simple expansion does not of course affect the transmission.

To ensure that the advantages of this approach are fully realised, the system must be carefully engineered. The edges of the slots should be defined with knife edges, bevelled on the side away from the source to minimise reflections. The slots themselves should be wide enough to allow ample clearance for the beam. The disc should be rigidly mounted and balanced, so that there is no vibration when it is in motion. If several discs are to be used in order to obtain a variety of transmission factors, it must be possible to interchange these without difficulty, but achieving good positional repeatability, so that the transmission factor does not have to be re-measured on each occasion.

The basic assumption made in the application of sectored discs is that the system obeys the photoelectric equivalent of Talbot's law, i.e. that the mean signal obtained from a strong source of spectral radiance L_λ observed through the rotating disc which transmits for a fraction β of the time is exactly the same as that obtained from a weaker source of radiance βL_λ observed continuously. Clearly the time constant of the output signal should be large enough to smooth out the steps in the chopped signal. This implies that the detector is linear in the time regime around the period of rotation of the disc. Much slower changes should not therefore be important. The ratio of the signals from a single source viewed through the rotating disc should not depend upon the speed of the disc if the output time constant remains long enough, and should be equal to the transmission of the disc determined from its geometry. It may prove necessary to restrict the radiance at the detector to satisfy these requirements.

5.8.2 Neutral density filters

At first sight, neutral density filters might not appear very useful for the functions under discussion here. The transmission at a given wavelength cannot be measured independently, but only by optical methods. It is not stable at the levels of accuracy required here (0.01-0.1 %), varying mildly with temperature and drifting slowly with time. However, neutral density filters do have a number of advantages which have lead to their use in this area. They are simple, inexpensive and readily available. The transmission is independent of the intensity of the incident radiation at the levels normally encountered in radiation thermometry. And, while the transmission is unstable, it is sufficiently stable that accurate measurements may be made in the short term. In particular, they are of value for establishing the linearity of a detector and determining the corrections, if any, for departures from this.

A neutral density filter with attenuation A_0, where attenuation is defined as the inverse of the transmission, reduces an initial radiance L_λ to $L\lambda/A_0$ measured at the detector. The apparent attenuation of the filter measured from the detector signals is

$$A = \frac{I_2}{I_1}$$

$$= \frac{a_1 L\lambda - a_2 L\lambda^2}{a_1 \frac{L\lambda}{A_0} - a_2 \left(\frac{L\lambda}{A_0}\right)^2}$$

Rearranging this gives

$$A = A_0 - \frac{a_2}{a_1^2}(A_0 - 1)L_l\lambda \tag{5.29}$$

However, the absolute value of l_λ is not known, and we need to write this equation in terms of one of the measured quantities. Taking the output current I_2, the larger signal obtained with the filter not in the beam, we obtain

$$A = A_0 - \frac{a_2}{a_1^2}(A_0 - 1)I_2 - \ldots \qquad (5.30)$$

Thus if we plot the apparent attenuation of the filter as a function of the larger output I_2, we should obtain a straight line whose intercept at the vertical axis gives the true attenuation A_0, and whose slope can then be used to calculate the non-linearity factor (a_2/a_1^2). Any measured signal current I can then be corrected to a linearised value I_0 using

$$I_0 = I\left(1 + \frac{a_2}{a_1^2}I\right) \qquad (5.31)$$

It should be emphasised that only the first-order terms have been retained in this analysis. The need to allow for higher-order contributions will be indicated by the non-linearity of the plot of A against I_2. While it is tedious but not difficult to derive the appropriate terms using an extension of the methods indicated above, the improvement in linearity then becomes limited by the uncertainties in the assumptions made and in the measured coefficients.

5.8.3   Multiple aperture techniques

In another simple technique, a set of apertures in a plate is placed in the beam entering the radiation thermometer in such a way that it is relatively uniformly illuminated. Each aperture has a separate shutter so that the radiation thermometer may receive light through any one of the apertures singly or in combination. The signal observed when a combination of apertures is open should be equal to the sum of the signals obtained separately from the appropriate apertures. It is generally convenient to make the apertures of equal area, or of some integral multiple of this, although the signals observed are not likely to retain this simple relationship. As with the neutral density filters, a correction for small deviations from linearity may be derived.

There are two major limitations to this approach. First, the beams traversing different apertures do not in general possess the same paths through the radiation thermometer, nor illuminate exactly the same area of the photodetector. The addition of two beams is not therefore the same as the corresponding increase in the intensity at a given point on the detector. This may give rise to problems in the analysis. The second is more a practical problem. It is an inherent assumption in this technique that the state of one aperture, i.e., whether open or closed, should not affect the contribution to the observed signal from another aperture. There are however a number of ways in which an interaction may be produced. In particular, if the source is not a black-body, reflections from closed shutters may be re-reflected by the source through open apertures. Alternatively, if the radiation thermometer itself reflects back part of the incident radiation, this may be in part returned by those shutters which are closed. To reduce these effects, the plate and shutter assemblies should be painted matt black. It is possible, using diffusers or integrating spheres to produce a more complex system which behaves in a more nearly ideal manner, but the overall transmission of such a system is extraordinarily low.

If the signals through apertures 1 and 2 separately are I_1 and I_2, and together I_3, then the correction factor may be obtained from

$$I_1 + I_2 - I_3 = 2\frac{a_2}{a_1^2}I_1I_2 \tag{5.32}$$

5.8.4 Beam conjoiners

A much more sophisticated and useful form of multiple aperture systems is provided by devices known generally as beam combiners or conjoiners. These take separate beams from one or more sources and combine them to give a beam which fills the field of view of the pyrometer. The advantage of a single source device is that the system is simpler to set up, and that it is easier to keep a single source stable to the accuracy required over the period of a set of measurements. On the other hand, the multiple source devices are more versatile, and allow a wide range of radiance ratios to be established by setting the lamps at different temperatures.

CHAPTER 6

Series expansion analytical technique

6.1 INTRODUCTION

A direct implementation of Planck's radiation law requires a thermometer with a narrow spectral bandwidth and a detector with linear response. By careful measurement of the spectral transmission of the optics and the detector spectral response very low uncertainties can be achieved, as exemplified by the so-called 'primary radiation thermometers' used in establishing accurate values of fixed-point (and other) temperatures or realising the International Temperature Scale. As shown in Chapter 4, the response of the radiation thermometer can be written as an integral over wavelength of the optical transmission, the detector response and the Planck function as in (6.4). The quantity k is a calibration factor which depends on the absolute detector response but must be independent of the thermal radiation intensity and wavelength. Where k is known the instrument can be used absolutely, but more usually it is calibrated at a known temperature (fixed-point) so that k can be eliminated by taking radiance ratios.

There are in fact two distinct aspects to the analysis of the data obtained; the results must be converted into values of temperature, and in addition a detailed assessment of the contributions to the total uncertainty from the various sources of random and systematic errors is necessary.

Fortunately, narrow-band radiation thermometry has the notable advantage that, provided a suitable wavelength is used for the temperature range of interest, the signal varies rapidly with the source temperature T. Taking the Wien approximation

$$M(\lambda, T) = c_1 \lambda^{-5} \exp(-c_2/\lambda T) \tag{6.1}$$

we have a simple expression for the radiance at two temperatures T and T_0,

$$\ln Q = \ln \frac{I(T)}{I(T_0)} = \frac{c_2}{\lambda'} \left(\frac{1}{T_0} - \frac{1}{T} \right) \tag{6.2}$$

where λ' characterises the operating wavelength of the thermometer and, neglecting the optical bandwidth of the instrument, the errors produced are not excessive.

Analytical techniques to convert signal to temperature were essential in the days before widespread availability of processing power. Nowadays it is a straightforward matter to numerically solve the Planck equation. Since the analytical approach is still used, we will consider it in Appendix B. The other use for evaluation of uncertainties will be considered here. We shall illustrate the significance of an analytical technique developed by the author by evaluation of a model radiation thermometer. This hypothetical instrument contains an interference filter with a transmission curve $t(\lambda)$ which is Gaussian in shape, has a centre wavelength of 660 nm and a bandwidth of 30 nm. The detector will be taken to be a photomultiplier with a S-20 cathode.

A reasonable operating range for such an instrument would be from 800 °C to 2400 °C. If we were to determine the ratio of the signal obtained at the platinum point (2040 °C) to that at the gold point (1064.18 °C) we should obtain an answer of 274.61. If we substitute this number into (6.2) and take the nominal centre wavelength (660 nm) of the filter for λ', the calculated temperature for the platinum point using the gold point as a reference is in error by only 1.3 °C. Considering the limited extent of the data employed in obtaining this result, the error is clearly not excessive.

Moreover, (6.2) is quite adequate for estimating the effects of error and uncertainty in the values of Q, λ' and T_0 on the calculated temperature T. Differentiation gives

$$\frac{\delta T}{T} = \left(\frac{c_2}{\lambda T}\right)^{-1} \frac{\delta Q}{Q} + \left(\frac{T}{T_0} - 1\right) \frac{\delta \lambda'}{\lambda'} + \left(\frac{T}{T_0}\right) \frac{\delta T_0}{T_0} \tag{6.3}$$

When δQ, $\delta \lambda'$ and δT_0 represent independent systematic effects or corrections, they may be combined directly with this equation, with due regard to their signs, of course. If however they indicate the magnitude of random errors, a root mean square (RMS) sum is normally taken, that is, the squares of the terms on the right-hand side are summed, and the square root taken to give $\delta T/T$.

The magnitude of the errors calculated with (6.3) are plotted separately in Figure 6.1 for our model radiation thermometer as a function of the temperature T with the gold point as the reference temperature T_0. In many cases T_0 will lie near the lower end of the operating range; the maximum values of the uncertainties, whether expressed in kelvins or as a percentage of T, will then occur at the maximum operating temperature. The permitted uncertainties in each variable should therefore be estimated at this point. For example, to limit the contribution to δT from wavelength uncertainty to 1 °C at 2400 °C, the wavelength of our model radiation thermometer must be known to within 0.25 nm. The absolute and percentage uncertainties in the calculated temperatures will then be less in all other parts of the operating range. It may be seen from Figure 6.1 that, unlike the other curves, the wavelength uncertainty goes to zero as T approaches T_0 so that the operating wavelength need be known much less accurately for small temperature differences.

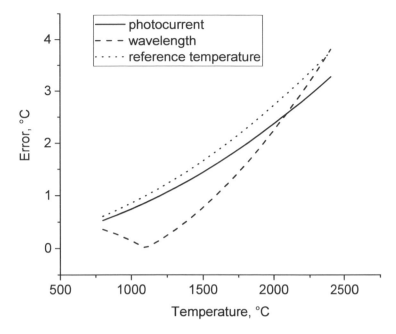

Figure 6.1: Effect of measurement errors and uncertainties on the calculated temperature T for a unit error in the signal ratio, the wavelength and the reference temperature.

In general, however, we cannot assume that the bandwidth of real radiation thermometers may be neglected, either in the calculation of temperature values or for the assessment of the more subtle uncertainties which originate in the assumptions made in modelling the behaviour of a real instrument. The measurement equation introduced in Chapter 4 may be written in the form

$$I(T) = k \int_0^\infty \epsilon(\lambda, T) P(\lambda, T) \tau(\lambda) t(\lambda) s(\lambda) d\lambda \qquad (6.4)$$

where the first two terms in the integral describe the characteristics of the source, $\tau(\lambda)$ the spectral transmission of optical components such as lenses, windows or neutral density filters, $t(\lambda)$ the spectral transmission of the colour or interference filter, and $s(\lambda)$ the spectral responsivity of the detector. The constant k contains a number of multiplying constants and the geometrical factors which determine the throughput of the thermometer. It is assumed either that the response does not depend upon the optical path through the instrument, or that the angular distribution of the radiation entering remains constant in all measurements. It should be noted that there are some situations where this assumption is not valid. In addition, it is assumed that corrections have been made for systematic errors such as the non-linearity of the detector and the so-called 'size-of-source effect'.

As it is extremely difficult to measure k to much better than 1 %, photoelectric radiation thermometers usually measure the unknown temperature T by reference to a known or defined temperature T_0. In the ITS-90, all measurements are directly or indirectly traceable to one of the freezing point of gold, which is assigned a value of 1337.33 K, or silver, which is assigned a value of 1234.93 K, or copper, which is assigned a value of 1357.77 K. Two techniques are available for the comparison; if the detection system is linear, or may be easily corrected to give a linear response, then T may be obtained from the ratio of the measured signal $I(T)$ to that from the reference T_0:

$$\frac{I(T)}{I(T_0)} = \frac{\int_0^\infty \epsilon(\lambda, T)P(\lambda, T)\tau(\lambda)t(\lambda)s(\lambda)d\lambda}{\int_0^\infty \epsilon(\lambda, T_0)P(\lambda, T_0)\tau(\lambda)t(\lambda)s(\lambda)d\lambda} \tag{6.5}$$

If the detector is non-linear or unstable, then a null detection system may be employed, in which an optical attenuator or neutral filter of known transmission $N(\lambda)$ is placed in front of the source at the higher temperature, and either $N(\lambda)$ or T_0 varied until the signals from both sources are equal (or nearly so). For example, if $T > T_0$

$$\int_0^\infty \epsilon(\lambda, T)P(\lambda, T)N(\lambda)\tau(\lambda)t(\lambda)s(\lambda)d\lambda$$
$$= \int_0^\infty \epsilon(\lambda, T_0)P(\lambda, T_0)\tau(\lambda)t(\lambda)s(\lambda)d\lambda \tag{6.6}$$

If $N(\lambda)$ is independent of wavelength, that is, the filter is completely neutral in the range covered by the interference filter, this term may be taken out of the integral on the left-hand side, and (6.5) and (6.6) are essentially identical, with $N(\lambda) \equiv I(T)/I(T_0)$. The transmittance τ (or the reflectance for any mirrors in the optical system) is designed to be at a maximum in the operating wavelength range of the thermometer, and therefore reasonably independent of wavelength. It will also be removed from the integrals in order to simplify the expressions to be developed in the following pages. It will be demonstrated later when we consider the dependence of radiance temperature on bandwidth that moderate departures from neutral behaviour may be accurately corrected with the techniques to be described.

If sufficient information is known about the functions $\tau(\lambda)$, $t(\lambda)$ and $s(\lambda)$, or equivalently the spectral responsivity $\Phi(\lambda)$ of the complete instrument, where

$$\Phi(\lambda) = \tau(\lambda)t(\lambda)s(\lambda)$$

then (6.5) and (6.6) may be solved for T using numerical techniques. The procedure involves a rapidly converging iterative method. The first step is to estimate λ' from the measured curve for $t(\lambda)$ or $\Phi(\lambda)$. With this and the measured signal ratio Q, a first estimate T_1 for the required temperature is obtained from (6.2). When substituted into (6.5), a corresponding signal ratio Q_1 is calculated from the integrals which is of course generally incorrect. However, a corrected temperature T_2 can be derived from this using an iterative equation obtained by differentiating (6.2),

$$T_{n+1} = T_n \left[1 + \frac{\lambda' T_n}{c_2} \left(1 - \frac{Q_n}{Q} \right) \right] \tag{6.7}$$

The process is repeated with T_2 the new estimate until consecutive values of the temperature agree with the accuracy required. It should be noted, first, that only a rough estimate is required for λ', as errors in this make little difference to the rate of convergence or the final result, and, secondly, that the integral in the denominator of (6.5) need only be calculated once, as it does not change between iterations.

6.2 THE SERIES EXPANSION TECHNIQUE

The highest accuracy measurement requires detailed assessment of uncertainties, that is, an estimation of the effect of random and systematic measurement errors upon the final uncertainty in the result. While numerical integration of (6.4) is fairly straightforward to determine the result, this approach is not suitable for detailed error analysis. Before the advent of readily available processing power the concept of an 'effective wavelength' was in widespread use to simplify the determination of a temperature value from a radiation thermometer signal. As shown in Appendix B, by using the Wien approximation in the derivation, the effective wavelength of a radiation thermometer is given by an integral equation

$$\frac{1}{\lambda_e} = \frac{\int_0^\infty \lambda^{-1} P(\lambda, T) t(\lambda) s(\lambda) d\lambda}{\int_0^\infty P(\lambda, T) t(\lambda) s(\lambda) d\lambda} \tag{6.8}$$

By solving this equation, the effective wavelength of a narrow-band radiation thermometer is found to be closely approximated by the equation

$$\frac{1}{\lambda_e} = a_0 + \frac{a_1}{T} \tag{6.9}$$

and the mean effective wavelength used to calculate a temperature is given by

$$\frac{1}{\lambda_m} = a_0 + \frac{a_1}{2} \left(\frac{1}{T_0} - \frac{1}{T} \right)$$
$$= \frac{1}{2} \left(\frac{1}{\lambda_{e0}} - \frac{1}{\lambda_e} \right) \tag{6.10}$$

The effective wavelength can be used in error analysis, but lacks physical significance, and it is therefore difficult to use it to gain any insight into the measurement problem. Repeated computer evaluation of the integrals can be used to investigate the effects of changes in particular parameters, but interactions between variables may be missed.

As an alternative, the series expansion technique developed here overcomes many of these problems, and is particularly intended to provide a rapid and versatile way of analysing the effects of small errors upon the calculated temperature. If we look at the various terms within the integral in (6.4), as shown in Figure 6.2, it is clear that the range of significant values is determined by the bandwidth of the colour filter. If $\tau(\lambda)$ is very narrow, then a good approximation to the signal is

$$I_0(T) = k\epsilon(\lambda_0, T) P(\lambda_0, T) s(\lambda_0) \int_0^\infty t(\lambda) d\lambda \tag{6.11}$$

that is, the variables, apart from the transmission $t(\lambda)$ of the colour filter, are assumed to be constant across the pass-band of the filter, and are replaced by their value at the centre wavelength λ_0. The integrated transmission of the filter is represented by

$$\int_0^\infty t(\lambda)d\lambda = a_0 t_0 \Delta\lambda \tag{6.12}$$

where a_0 is a measure of the shape of $t(\lambda)$, t_0 is an abbreviation for the transmission of the filter at λ_0, and $\Delta\lambda$ is the FWHM (full width at half maximum) bandwidth of the filter.

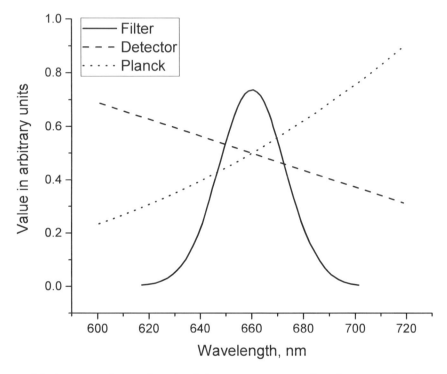

Figure 6.2: Variation with wavelength of the main terms within the integrals: the transmission of the filter $\tau(\lambda)$, the Planck function for a source temperature of 1500 °C, and the spectral responsivity of an S-20 photomultiplier.

The question which then arises is how to allow for the finite bandwidth of real radiation thermometers. We may regard (6.11) as the zeroth-order term in the Taylor series expansion of the product within the integral, and this suggests a means of calculating the correction $C(T)$ to $I_0(T)$ to give an accurate representation of $I(T)$ in the general case. That is,

$$C(T) = \frac{I(T)}{I_0(T)} - \int_0^\infty \frac{\epsilon(\lambda,T)}{\epsilon(\lambda_0,T_0)} \frac{P(\lambda,T)}{P(\lambda_0,T_0)} \frac{s(\lambda)}{s(\lambda_0)} \frac{t(\lambda)}{a_0 t_0 \Delta\lambda} d\lambda$$

Each term apart from $t(\lambda)$ is then expanded as a modified Taylor series about the centre wavelength λ_0, for example

$$\frac{s(\lambda)}{s(\lambda_0)} = 1 + s_1 \left(\frac{\lambda - \lambda_0}{\lambda_0}\right) + s_2 \left(\frac{\lambda - \lambda_0}{\lambda_0}\right)^2 + \cdots \qquad (6.13)$$

where the (non-dimensional) coefficients s_n are related to the derivatives of $s(\lambda)$ by

$$s_n = \left|\frac{d^n s(\lambda)}{d\lambda^n}\right|_{\lambda_0} \frac{\lambda_0^n}{n! s_0} \qquad (6.14)$$

Multiplying the series together and collecting the terms in $(\lambda - \lambda_0)^n$ enables $C(T)$ to be written as a series of sums multiplied by the integrals

$$\int_0^\infty (\lambda - \lambda_0)^n t(\lambda) d\lambda = a_n t_0 \Delta \lambda^{n+1} \qquad (6.15)$$

The coefficients a_n are the moments of the distribution $t(\lambda)$ about λ_0. If we define λ_0 as the mean wavelength, some simplification is achieved as a_1 is then zero by definition. The expression for the correction factor obtained from (6.13) may then be written

$$C(T) = 1 + A_2 \left(\frac{\Delta \lambda}{\lambda_0}\right)^2 \Sigma_2 + A_3 \left(\frac{\Delta \lambda}{\lambda_0}\right)^3 \Sigma_3 + \cdots$$

where $A_n = a_n / a_0$. The values of A_n are obtained using (6.15) from the distribution $\Phi(\lambda)$ or $t(\lambda)$ which must be accurately measured in order to calculate λ_0. To give an idea of their magnitude and behaviour, the integrations have been carried out for some representative filters of simple form, and are listed in Table 6.1. It will be seen from these that, as expected, the odd order coefficients are zero for symmetrical shapes. The coefficients otherwise increase as the proportion of transmission in the wings increases, a square shape clearly being ideal in this respect.

Table 6.1: Values of A_n for simple filter shapes

Shape $t(\lambda)$	A_2	A_3	A_4
Square	0.0833	0	0.0125
Symmetric triangle	0.1667	0	0.0667
Asymmetric triangle	0.2222	0.0593	0.1185
Gaussian	0.1803	0	0.0976
DFP pyrometer	0.2505	0.1708	0.3668

The terms Σ_n are the sums of the modified Taylor coefficients and their cross-products of order n; for example

$$\Sigma_2 = \epsilon_2 + P_2 + s_2 + \epsilon_1 P_1 + \epsilon_1 s_1 + \epsilon_1 s_1 \qquad (6.16)$$

The coefficients P_n may be calculated from (6.14) and the Planck function $P(\lambda, T)$. In this chapter, we shall frequently employ the 'Planck parameter'

$$p = \frac{c_2}{\lambda T}$$

to simplify the working and the appearance of the more complex equations. The addition of subscripts to p refer to the temperature variable rather than the wavelength, which may in general be taken to be the mean wavelength λ_0. Using the 'Planck parameter', the exact form of the first two coefficients is

$$P_1 = \frac{p}{1 - e^{-p}} - 5$$
$$P_2 = \frac{1}{2} \left[\frac{p^2(1 + e^{-p})}{(1 - e^{-p})^2} - \frac{12p}{1 - e^{-p}} + 30 \right]$$

For many applications it is sufficient to use the simpler polynomials for P_n obtained from the Wien approximation:

$$P_1 = p - 5$$
$$P_2 = p^2/2 - 6p + 15$$
$$P_3 = p^3/6 - 7p^2/2 + 21p - 35$$
$$P_4 = p^4/24 - 4p^3/3 + 14p^2 - 56p + 70$$

The polynomials P_n are plotted in Figure 6.3. It will be seen that they increase rapidly above a value of p of 15, that is, at temperatures towards the lower end of the operating range to be expected from the detector used. Estimates of the remaining coefficients ϵ_n and s_n must be obtained from measured data or published values. Over the filter bandwidth, it is generally sufficient to represent both the emissivity and the responsivity of the detector by a quadratic in the wavelength. Within the correction terms, this means that only the first- and second-order terms are required, s_1 and s_2 for example, which simplifies the higher-order terms somewhat. For smoothly varying quantities, the magnitudes of these coefficients are typically less than 10, although they may of course be negative as well as positive. It is possible for the coefficients to exceed this value if the emissivity of the source is affected by a sharp absorption band, or the detector is being used in the long wavelength cut-off region, for example, where the responsivity falls very rapidly with wavelength. But under these conditions the thermometry results obtained will tend to be unreliable, and they should be avoided wherever possible.

A number of other simplifications are possible. For example, for calculating temperature values the effects of the source parameters $\epsilon(\lambda, R)$ and $P(\lambda, T)$ are often of greatest interest. The characteristics of the radiation thermometer may then be taken as fixed, and the thermometer wavelength function $\Phi(\lambda)$ used instead of $t(\lambda)$ in (6.15). The radiation thermometer characteristics are then completely

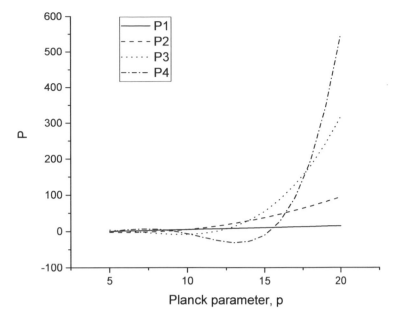

Figure 6.3: Values of the polynomials P_n in the range $p = 5$ to 20.

contained in the coefficients λ_0, $\Delta\lambda$ and the values of A_n and the summation terms Σ_n contain only the coefficients from the emissivity and the Planck distribution. The value obtained for the mean wavelength λ_0 is not, of course, exactly the same as that calculated from the filter transmission curve $t(\lambda)$ and the two should not be regarded as numerically interchangeable.

6.2.1   Extent of applications of the technique

To summarise the usage of the series expansion technique, we note that the integrals occurring within equations of the types shown in (6.5) and (6.6) may be replaced by the product of an idealised signal $I_0(T)$ and a correction factor $C(T)$, which consists of a power series in the fractional bandwidth $\Delta\lambda/\lambda_0$ of the colour filter. As the measurement equations involve the ratio of two signals, the factors s_0, a_0, t_0 and $\Delta\lambda$, as well as the geometrical constant k, in the expressions for $I_0(T)$ and $I_0(T_0)$ cancel out. The result is a ratio of simple Planck functions at a fixed wavelength λ_0 multiplied by the ratio of two correction factors, each of which is close to unity.

Clearly the rapid convergence of the series $C(T)$ is a critical point. If the method is to be useful in practice, we need to avoid the evaluation of the more complex higher-order terms, which may contain coefficients of unknown magnitude. It is evident from Figure 6.3 that the problems are likely to be greatest at high values of p, that is, at low temperatures, where the polynomials P_n increase rapidly. In this region, however, the effect of the emissivity and spectral responsivity coefficients will normally be small, so that we may judge the rate of convergence from the P_n

alone. Under these conditions

$$\Sigma_n < \frac{(p-5)^n}{n!}$$

At high temperatures, the P_n are small, so that it is sufficient to ensure that none of the other coefficients is excessively large. It should be noted in this context that the P_n are calculable and therefore in principle free from error, while the other coefficients are derived from experimental measurements and their uncertainties increase rapidly with increasing n.

Rapid convergence may of course be assured by using filters with small fractional bandwidths. The interference filters in many primary radiation thermometers have bandwidths of around 10 nm, and this is sufficiently small that $C(T)$ may be limited to two terms, that is, $n = 2$. However, much of modern primary radiation thermometry has been directed towards extending the range of accurate measurements down to as low a temperature as possible, and broader filters represent one simple means of achieving this aim. Even then, it is possible to improve the rate of convergence by selecting filters with a symmetrical transmission curve and a sharp cut-off on either side. It will be seen from Table 6.1 that these will have the smallest values for the A_n factors, with near-zero values for odd n.

In selecting a filter there is a fundamental difference between the considerations which determine the bandwidth and those relevant to the shape of $t(\lambda)$ given by the A_n. As stated above, the bandwidth may be increased to enable lower temperatures to be measured with a given wavelength and detector. Two factors set the limit for this; as the bandwidth is increased, the higher-order terms in $C(T)$ become increasingly important, and in general these contain coefficients ϵ_n and s_n which are inaccurately known for high n. In addition, the uncertainty in the measurement of λ_0 must remain unchanged, and this becomes more difficult when the bandwidth is large. On the other hand, the effects of out-of-band transmission, that is, in the wings of $t(\lambda)$ or from insufficiently blocked sidebands, decrease as the bandwidth increases. As a result, there is a broad range of near-optimum fractional bandwidths, from around 0.01 up to 0.05 in the visible and near infrared.

The effects which limit the measurement accuracy obtained are minimised by selecting filters with low values for the A_n coefficients, that is, whose transmission curve is symmetrical and as nearly square as possible, and which have out-of-band transmission falling to a few parts per million or less. The presence of quite weak sidebands is often sufficient to degrade the overall performance of the thermometer. The set A_n therefore acts as a figure of merit for the filter and may be used to separate good filters from bad, while the bandwidth may be selected to suit a particular application.

In general though, it is worthwhile to establish the maximum fractional bandwidth for which the technique remains useful down to the minimum temperature of interest. In fact the criterion of acceptable behaviour depends to some extent on the particular application. For example, if it is to be used for the calculation of temperatures from precise measurements, the uncertainty in $C(T)$ from truncation and from errors in any of the constituent terms must be less than the measurement uncertainty in the radiance ratio. For primary thermometry this may be taken at about one part in 10^4, but for secondary calibrations and industrial measurements it would of course be considerably larger. For the reasons suggested in the previous paragraphs, it is undesirable to calculate any term greater than the fourth in the series ($n = 4$). Assuming that this last term can be estimated with an accuracy of about 10 %, then the fractional bandwidth of a typical interference filter should be limited to about 0.05 for a maximum value of p of 20.

On the other hand, for the investigation of small systematic errors it is desirable that only the second term need be considered, and the remainder neglected. Fortunately, it is usually necessary to calculate the uncertainty to within 5-10 %, so that the remaining terms should be less than the $n = 2$ term by a factor of 20 or more. For filters with moderately symmetrical transmission curves, this requirement corresponds to a fractional bandwidth of about 0.15 for a maximum value of p equal to 20. It is apparent that, for this application, a greater range of filter bandwidths may be usefully included.

It should be stressed that the series expansion technique is essentially a numerical approximation method, and at each stage it is advisable to check, using whatever information is at hand, that the assumptions made about the relative magnitudes of the terms are sensible. Familiarity obtained in this way quickly enables the reliability of the results obtained to be assessed.

6.3 EXAMPLES OF APPLICATIONS

6.3.1 Calculation of temperatures from signal ratios

We will consider an instrument with a 10 nm-wide filter with a mean wavelength of 655 nm with an S-20 photomultiplier, to operate over the temperature range from 800 °C to 2300 °C and above. We will calculate $C(T)$ over this range for such an instrument, assuming that the sources may be treated as black-bodies.

It has been shown by measurements taken at the National Physical Laboratory over many years that the spectral responsivity of S-20 photocathodes may be described by a linear equation over a range of more than 100 nm about a mean wavelength of 655 nm, that is, we may take the coefficient s_2 to be zero. The average for s_1 was found to be -4.16, with a total spread of about 20 % around this value. The effect of the uncertainty in s_1 will be covered in a later section. Assuming that the shape of the filter is Gaussian – in fact the transmission curves of most multilayer interference filters are squarer than this – we may readily calculate the contributions to $C(T)$ from the various terms. With the parameters given, the fourth term in the series (i.e., with $n = 4$) is never greater than a few parts in a million, and, as the filter is symmetrical, the third term is also negligible. $C(T)$ is therefore accurately given by

$$C(T) = 1 + A_2 \left(\frac{\Delta\lambda}{\lambda_0}\right)^2 (P_2 + s_1 P_1)$$

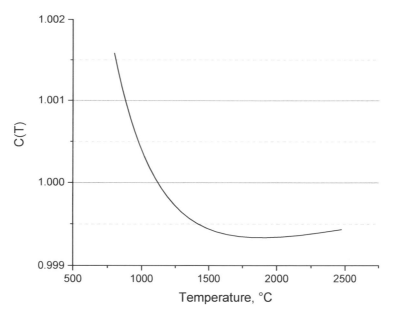

Figure 6.4: The correction factor $C(T)$ for a radiation thermometer with a bandwidth of 10 nm and an S-20 photomultiplier detector.

and is plotted in Figure 6.4 for the data given above. To determine the temperature T from the signal ratio with respect to a reference source at T_0 it is only necessary to solve

$$\frac{I(T)}{I(T_0)} = \frac{P(\lambda_0, T)}{P(\lambda_0, T_0)} \cdot \frac{C(T)}{C(T_0)}$$

with an iterative procedure which should converge rapidly. To meet the criterion for the calculation of temperature values, the only measured coefficient, s_1, needs to be determined with an uncertainty of about 10 %. For measurements between the gold point and higher temperatures, the value of s_1 given above is adequate even for the most accurate calculations. The results imply that changes in the responsivity of the photocathode from ageing or temperature variations are unlikely to cause significant errors with this class of instrument. However, if the operating range were extended to lower temperatures, then s_1 would need to be measured more carefully, and it is possible that temperature control of the photomultiplier would be necessary. Cooling of the photocathode would also reduce the background noise component from thermionic emission.

6.3.2 Generalised effective wavelengths

Effective wavelengths are generally calculated for black-body sources. In this section we shall attempt to see to what extent the concept remains meaningful when the sources are not black-bodies. The inconvenience of determining and specifying the emissivity of a standard source, such as a tungsten strip filament standard lamp, is avoided by calibrating it in terms of the spectral radiance temperature, defined by

$$\epsilon(\lambda, T)P(\lambda, T) = P(\lambda, T_R) \tag{6.17}$$

rather than the true temperature of the surface. A simplified version of this equation, suitable for use within the correction terms, may be derived from the Wien approximation

$$p_R = p - \ln \epsilon_0 \tag{6.18}$$

where ϵ_0 is an abbreviated form of $\epsilon(\lambda_0, T)$. The general definition of the mean effective wavelength between two radiance temperatures T_{R1} and T_{R2} is given from (B.10) by

$$\frac{I(T_{R1})}{I(T_{R2})} = \frac{P(\lambda_m, T_{R1})}{P(\lambda_m, T_{R2})}$$

$$= \frac{\int_0^\infty \epsilon(\lambda, T_1)P(\lambda, T_1)t(\lambda)s(\lambda)d\lambda}{\int_0^\infty \epsilon(\lambda, T_2)P(\lambda, T_2)t(\lambda)s(\lambda)d\lambda}$$

where T_1 and T_2 are the true temperatures of the two sources.

Substitution for the integrals using the series expansion method gives

$$\frac{P(\lambda_m, T_{R1})}{P(\lambda_m, T_{R2})} = \frac{\epsilon(\lambda_0, T_1)P(\lambda_0, T_1)}{\epsilon(\lambda_0, T_2)P(\lambda_0, T_2)} \cdot \frac{C(T_1)}{C(T_2)}$$

Introducing the radiance temperatures from (6.17),

$$\frac{P(\lambda_m, T_{R1})}{P(\lambda_m, T_{R2})} = \frac{P(\lambda_0, T_{R1})}{P(\lambda_0, T_{R2})} \cdot \frac{C(T_1)}{C(T_2)}$$

Replacing the Planck function by the Wien approximation with the attendant correction

$$P(\lambda, T) = \frac{W(\lambda, T)}{1 - e^{-p}}$$

and, noting that the correction terms cancel to high accuracy since λ_m is close to λ_0, taking logs gives

$$\frac{c_2}{\lambda_m}\left(\frac{1}{T_{R2}} - \frac{1}{T_{R1}}\right) = \frac{c_2}{\lambda_0}\left(\frac{1}{T_{R2}} - \frac{1}{T_{R1}}\right) + \ln C(T_1) - \ln C(T_2)$$

To calculate the limiting effective wavelength from this equation, we shall divide throughout by

$$c_2\left(\frac{1}{T_{R2}} - \frac{1}{T_{R1}}\right) \equiv \lambda_0(p_{R2} - p_{R1})$$

and take the limit as $T_{R2} \to T_{R1}$. Substituting for the correction factors and expanding the log terms gives

$$\frac{1}{\lambda_m} = \frac{1}{\lambda_0} \left\{ 1 + \frac{1}{p_{R2} - p_{R1}} \left[A_2 \left(\frac{\Delta\lambda}{\lambda_0} \right)^2 (\Sigma_2(T_1) - \Sigma_2(T_2)) + \ldots \right] \right\}$$

For a first approximation we terminate the series at the second term, i.e., to that shown. We may then substitute for the Σ_2 terms from (6.16). To simplify the expression further two assumptions are necessary; first, that the emissivities $\epsilon(\lambda, T)$ for the two sources are the same, and, secondly, that the temperature dependence of the emissivity coefficients may be neglected. While the latter is of little significance, since we shall be taking the limit as the source temperatures approach each other, the first represents a major limitation to the application of the effective wavelength concept. In this case we have, however, after some manipulation,

$$\frac{1}{\lambda_m} = \frac{1}{\lambda_0} \left\{ 1 + \frac{p_1 - p_2}{p_{R2} - p_{R1}} A_2 \left(\frac{\Delta\lambda}{\lambda_0} \right)^2 [(p_1 + p_2)/2 - 6 + s_1 + \epsilon_1] \right\}$$

Substituting for p from (6.18)

$$\frac{1}{\lambda_m} = \frac{1}{\lambda_0} \left\{ 1 - A_2 \left(\frac{\Delta\lambda}{\lambda_0} \right)^2 [(p_1 + p_2)/2 + \ln \epsilon_0 - 6 + s_1 + \epsilon_1] \right\}$$

We now allow T_{R2} to approach T_{R1} (and hence $p_{R2} \to p_{R1}$) to obtain the limiting effective wavelength λ_e,

$$\frac{1}{\lambda_e} = \frac{1}{\lambda_0} \left\{ 1 - A_2 \left(\frac{\Delta\lambda}{\lambda_0} \right)^2 [p_R + \ln \epsilon_0 - 6 + s_1 + \epsilon_1] \right\} \tag{6.19}$$

dropping the unrequired subscript indicating the temperature source. The significance of this derivation is that it indicates that the concept of an effective wavelength can be applied to sources other than black-bodies, but only under the very restrictive condition that the emissivity characteristics of the two sources are closely similar. In practice, as the ultimate reference source is likely a black-body, this condition limits the precise application of effective wavelengths to other black-bodies.

The limiting effective wavelength for a black-body source at a temperature T_R may easily be found by eliminating the emissivity terms

$$\frac{1}{\lambda_e^B} = \frac{1}{\lambda_0} \left\{ 1 - A_2 \left(\frac{\Delta\lambda}{\lambda_0} \right)^2 \left[\frac{c_2}{\lambda_0 T_R} - 6 + s_1 \right] \right\} \tag{6.20}$$

Although this equation contains a number of approximations, it provides a very simple method for estimating the effective wavelength from readily measured parameters of the system, the main effort required being in the determination of the filter transmission curve $t(\lambda)$. Values calculated in this way for our model radiation

Table 6.2: Comparison of limiting effective wavelengths for our model radiation thermometer calculated by different methods

	By integration		By series expansion	
$T/°C$	λ_W (B.6)	λ_P (B.5)	Approximate (6.20)	+ Extra terms (6.21)
800	662.446	662.446	662.506	662.446
1000	661.679	661.679	661.716	661.680
1200	661.119	661.119	661.142	661.120
1400	660.692	660.690	660.706	660.692
1600	660.355	660.349	660.364	660.355
1800	660.082	660.064	660.087	660.082
2000	659.857	659.812	659.860	659.857
2200	659.669	659.570	659.669	659.668
2400	659.508	659.318	659.507	659.507

thermometer are shown in Table 6.2, together with those determined by integration in Appendix B.

Plots of the reciprocal of the effective wavelength against $1/T$ are shown in Figure 6.5 for similar filters with a range of fractional bandwidths, again assuming an S-20 photomultiplier response with s_1 coefficient of -4.16 based on measurements of the detector spectral response. As expected from (6.20), the plots are linear with a slope which increases with the square of the fractional bandwidth. The lines intersect at a common point where the effective wavelength is equal to λ_0, and

$$\frac{c_2}{\lambda_0 T_R} = 6 - s1$$

The difference $\lambda_e - \lambda_e^B$ obtained from (6.19) and (6.20) is independent of the radiance temperature, that is, it appears as a fixed offset in the effective wavelength. The corresponding temperature error for our model radiation thermometer when used to compare a tungsten strip lamp at the platinum point with a similar reference lamp at the gold point amounts to $0.33\,\text{K}$. It should be emphasised that these equations do not allow the errors which occur when the effective wavelength calculated for black body sources is used in the comparison of a black-body and a source of lower emissivity. This problem, as stated above, cannot be tackled directly with the effective wavelength method; it has been solved with the series expansion technique [19].

6.3.3 Calculation of second-order terms

While (6.20) gives a good approximation to the limiting effective wavelength obtained by integration, especially at high temperatures, it is an interesting exercise to calculate the second-order terms which account for the differences. Under the

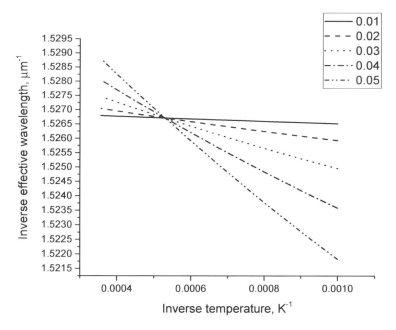

Figure 6.5: Variation of inverse effective wavelength in μm^{-1} *for different fractional band-widths.*

conditions assumed in the previous section, this turns out to be relatively simple, as the sources are treated as black-bodies, and the spectral responsivity term is represented fully by the coefficient s_1. A further simplification of the method in this particular case is possible; the process of division by $(p_{R2} - p_{R1})$ followed by taking the limit as $p_{R2} \to p_{R1}$ is equivalent to taking the differential of $-\ln C(T)$ with respect to p_R. Only the Planck polynomials P_n contribute to the result, as these alone contain p_R.

If $C(T)$ is represented by

$$C(T) = 1 + \Xi_2 + \Xi_3 + \Xi_4 + \ldots$$

where Ξ_n is the nth term in the series expansion, then

$$\ln C(T) = (\Xi_2 + \Xi_3 + \Xi_4 \ldots) - \frac{1}{2}(\Xi_2 + \Xi_3 + \Xi_4 \ldots)^2 + \ldots$$

Rough numerical estimates of the magnitudes of these terms indicate that only Ξ_2, Ξ_4 and Ξ_2^2 are of any significance under the circumstances of this calculation, as the filter transmission curve is symmetrical. However, we will also derive the expression for Ξ_3 for the sake of completeness. Combining the terms in Ξ_2

$$\frac{d}{dp_R}\left(\Xi_2 - \frac{\Xi_2^2}{2}\right) = \frac{d\Xi_2}{dp_R}(1 - \Xi_2)$$

$$= A_2\left(\frac{\Delta\lambda}{\lambda_0}\right)(p_R - 6 + s_1)\left[1 - A_2\left(\frac{\Delta\lambda}{\lambda_0}\right)^2(P_2 + s_1 P_1)\right]$$

where

$$P_2 + s_1 P_1 = \frac{p_R^2}{2} - 6p_R + 15 + s_1(p_R - 5)$$

For the terms in Ξ_3 and Ξ_4,

$$\frac{d\Xi_3}{dp_R} = \frac{d}{dp_R}(P_3 + s_1 P_2)$$

$$= A_3 \left(\frac{\Delta\lambda}{\lambda_0}\right)^3 \left[\left(\frac{p_R^2}{2} - 7p_R + 21\right) + s_1(p_R - 6)\right]$$

and

$$\frac{d\Xi_4}{dp_R} = \frac{d}{dp_R}(P_4 + s_1 P_3)$$

$$= a_4 \left(\frac{\Delta\lambda}{\lambda_0}\right)^4 \left[\left(\frac{p_R^3}{6} - 4p_R^2 + 28p_R - 56\right) + s_1\left(\frac{p_R^2}{2} - 7p_R + 21\right)\right]$$

The complete expression for calculating λ_e is therefore

$$\frac{1}{\lambda_e} = \frac{1}{\lambda_0}\left[1 - \frac{d\Xi_2}{dp_R}(1 - \Xi_2) - \frac{d\Xi_4}{dp_R}\right] \tag{6.21}$$

Substitution of the values for A_2 and A_4 from Table 6.1 yields the corrected values for the effective wavelength shown in the last column of Table 6.2. It will be seen that they agree to one digit in the last figure with those obtained by integration based on using Wien's equation (6.8). However, the series expansion technique requires only five integrals (for the moments a_0 to a_4) to enable the effective wavelength to be calculated at any number of temperatures. Should the characteristics of the detector change, the same filter parameters remain valid, and only the value of s_1 need be changed.

6.3.4   Uncertainties from changes in the spectral response

In many radiation thermometry calculations, an assumption is made about the spectral behaviour of one or more of the components of the system. While it is usual to measure the spectral responsivity of the detector, for example, the measurements may be performed at room temperature while the detector is normally cooled in operation. Even if a correction is applied, there will be some associated uncertainty, arising from the imperfectly known temperature coefficient or variations in the final temperature. Similarly, it may be assumed that the transmission of optical components and neutral density filters is independent of wavelength, while in practice some departure from neutrality is present. In this section, we shall show how the series expansion technique allows the resulting uncertainties in the calculated temperature to be estimated, and the nature of the information required for this purpose.

We consider first the general case where a source at an unknown radiance temperature T_R corresponding to a real temperature T is measured by comparison to a reference black-body at T_0. We shall calculate the uncertainty in T_R arising from the possible measurement uncertainties or approximations in the function adopted for $s(\lambda)$. The measurement equation in this case may be represented by

$$\frac{I(T_R)}{I(T_0)} = \frac{\int_0^\infty \epsilon(\lambda, T) P(\lambda, T) s(\lambda) t(\lambda) d\lambda}{\int_0^\infty P(\lambda, T_0) s(\lambda) t(\lambda) d\lambda} \tag{6.22}$$

Following the usual procedure of replacing the integrals by $I_0(T).C(T)$, we obtain

$$\frac{I(T_R)}{I(T_0)} = \frac{P(\lambda_0, T_R)}{P(\lambda_0, T_0)} \left\{ 1 + A_2 \left(\frac{\Delta\lambda}{\lambda_0} \right)^2 [\Sigma_2(T) - \Sigma_2^B(T_0)] \right\} \tag{6.23}$$

where the superscript 'B' indicates that the term is evaluated under black-body conditions, that is, with the emissivity coefficients equal to zero. The Σ_2 terms may be expanded using (6.16) to give

$$\Sigma_2(T) - \Sigma_2^B(T_0) = [P_2(T) + s_2 + \epsilon_2 + P_1(T)\epsilon_1 + \epsilon_1 s_1] \\ - [P_2(T_0) + s_2 + P_1(T_0)s_1] \tag{6.24}$$

It will be seen that the curvature of the spectral responsivity, represented by s_2 cancels out from the equation. Although it reappears in the higher-order terms in $C(T)$, these are in general small, so that the temperature obtained is quite insensitive to variations in this coefficient. The major effect is therefore through the linear coefficient s_1. If the Planck functions in (6.23) are replaced with the Wien approximation, and the Σ_2 terms substituted from (6.24), differentiation with respect to s_1 allows the temperature error δT_R produced by a small change δs_1 to be derived:

$$\frac{\delta T_R}{T_R} = A_2 \left(\frac{\Delta\lambda}{\lambda_0} \right)^2 \frac{\delta s_1}{p_R} [(p_0 - p_R) + (\epsilon_1 + \ln \epsilon_0)] \tag{6.25}$$

The effect of the emissivity terms is in general small but not always negligible. A similar equation for black-body sources has been derived empirically by Bedford and Ma [20] by repeated computation of the integrals in (6.22). Their work was limited to a few simple models for the shape of the filter transmission curve $t(\lambda)$. In these cases, the numerical results agreed with those from the equation above, but the virtue of the latter is of course that it can be applied to any real filter whose transmission curve is known.

Equation (6.25) may also be employed to calculate the effect of the spectral coefficients, or changes in them, of any other component within the optical system of the thermometer. At an early stage in this chapter, we combined the spectral transmission of lenses, neutral density filters etc. into a factor $\tau(\lambda)$ which was absorbed into $s(\lambda)$. If we wish to estimate the effect of a change with wavelength in the transmission of a (nominally) neutral density filter, for example, we calculate the modified Taylor coefficients τ_1 and τ_2 from (6.14). As the effect of the second-order coefficients upon the temperature obtained is negligible, except for radiation thermometers with very broad, asymmetrical colour filters, the latter may be discarded. If the initial assumption was that the filter was completely neutral, then (6.25) may be used directly to calculate the temperature errors, with $\delta s_1 \equiv \tau_1$.

6.3.5 Dependence of radiance temperature upon bandwidth

The definition of a radiance or brightness temperature for a surface which is not a black-body source is an extremely convenient way of avoiding the need to know its emissivity, and particularly appropriate where the emissivity is stable and the true temperature of the surface is not required. The definition given in (6.17) equates the radiance of the surface at a temperature T with that from a black-body at the radiance temperature T_R at a given wavelength. While this may be used directly for applications requiring only moderate accuracy, for critical usage we need to take into account the finite bandwidth of the radiation thermometer. The correction is particularly important when the surface is used as a temperature standard, for example a tungsten strip filament lamp, which may be used with several radiation thermometers with different bandwidths.

For a real radiation thermometer, the definition given above is mathematically equivalent to

$$\int_0^\infty \epsilon(\lambda, T) P(\lambda, T) t(\lambda) s(\lambda) d\lambda = \int_0^\infty (\lambda, T_R^*) P(\lambda, T) t(\lambda) s(\lambda) d\lambda$$

where T_R^* is the radiance temperature determined with that radiation thermometer. Expanding the integrals with the series expansion technique gives

$$\epsilon(\lambda_0, T) P(\lambda_0, T) C(T) = P(\lambda_0, T_R^*) C(T_R^*) \tag{6.26}$$

If the bandwidth of the thermometer were negligible, then the correction terms would both be unity, and the radiance temperature T_R would be given by (6.17). Combining this with (6.26) yields

$$\frac{P(\lambda_0, T_R^*)}{P(\lambda_0, T_R)} = \frac{C(T)}{C(T_R^*)}$$

Since the difference between T_R and T_R^* should be small for narrow-band radiation thermometers, it is possible to make a number of valid approximations. First, the ratio of the Planck terms may be replaced by that of the Wien approximations with negligible error. Secondly, within the correction term on the right-hand side, T_R^* may be replaced by R_R and this may be related to the true temperature of the surface through (6.20). Expanding the correction factors in the usual way and taking logs gives

$$\frac{c_2}{\lambda_0}\left(\frac{1}{T_R} - \frac{1}{T_R'}\right) = A_2 \left(\frac{\Delta\lambda}{\lambda_0}\right)^2 [\Sigma_2(T) - \Sigma_2^B(T_R^*)]$$

where the superscript 'B' indicates that this term is written out with the emissivity coefficients equal to zero. If the Planck polynomials P_n are also expanded, and (6.18) is used to eliminate p, some elementary rearrangement results in the expression

$$\frac{T_R^* - T_R}{T_R} = A_2 \left(\frac{\Delta\lambda}{\lambda_0}\right)^2 \frac{1}{p_R}[\epsilon_2 + (p_R - 5)s_1 + \ln\epsilon_0(p_R - 6 + s_1 + \epsilon_1 + \ln\epsilon_0/2)]$$

(6.27)

The magnitude of the difference depends rather critically upon the characteristics of the thermometer, although it is in general quite small. For a tungsten strip filament lamp, with the true temperature of the tungsten at 2000 K, the emissivity coefficients of the lamp, which include the transmission of the envelope or windows, are typically at 660 nm:

$$\epsilon_0 = 0.425 : \epsilon_1 = -0.198 : \epsilon_2 = -0.023$$

so that the radiance temperature T_R is 1854 K (1580 °C) and p_R is 11.756. For the model radiation thermometer, the difference in radiance temperatures given by (6.27) is only 0.13 K.

CHAPTER 7

Multi-wavelength radiation thermometry

7.1 INTRODUCTION

As has been seen from the discussion earlier in this book, the major difficulty in obtaining accurate temperatures in radiation thermometry is the requirement for knowledge of the emissivity of the target surface. While monochromatic radiation thermometry can achieve very high accuracy when the surface characteristics are well known, in many applications this is unfortunately not the case. The determination of spectral emissivities for the wide range of materials of industrial and scientific interest is by no means a simple matter, especially at high temperatures, where the value obtained will depend significantly upon the roughness of the surface and the degree to which chemical changes have taken place. There has, therefore, been a great deal of interest in the performance of radiation thermometers designed to be of general application, and to avoid the need for an accurate value of the emissivity.

We consider first the suggestion that it is possible to estimate both the spectral emissivity and the true temperature from redundant information in the spectral distribution of thermal radiation from the surface. We will represent, in our analysis of this possibility, the signal I_i obtained within a narrow band around the wavelength λ_i by the Wien approximation

$$I_i = k_i \epsilon_i \lambda_i^{-5} \exp\left(\frac{c_2}{\lambda_i T}\right) \tag{7.1}$$

where ϵ_i represents the emissivity at the channel wavelength λ_i. Although ϵ_i is generally temperature dependent, it is unnecessary to invoke this dependence during the analysis, since we are dealing with the determination of a particular temperature, and it will be suppressed for the present. The constant k_i, containing a combination of geometrical factors, the responsivity of the detector and fundamental constants, may be regarded as known. It may be determined, for example, by calibration against a black-body source at a known temperature. While this equation is of course an approximation to the exact form, its use greatly simplifies the subsequent analysis, and does not invalidate the conclusions in any way, as corrections for the Wien approximation and for the finite bandwidth involved may be made with more than sufficient accuracy with the techniques described in previous chapters.

If we have no information about the spectral emissivity ϵ_i, we must assume a value in order to convert our measured signal I_i to an estimate of the true temperature T. If we treat the surface as a black-body, i.e. $\epsilon_i = 1$, the temperature obtained is known as the spectral radiance temperature T_m. It might be more reasonable, if we wish to minimise the possible uncertainty, to take a value of around 0.3, as this is near the centre of the range of possible values for the logarithm of ϵ_i. However, whichever assumption is made, the temperature calculated is likely to be seriously in error in the majority of applications. For example, if the true temperature of the surface is $1500\,\mathrm{K}$ ($1227\,^\circ\mathrm{C}$), and we use a narrow-band radiation thermometer with an operating wavelength of 655 nm, making the assumption that the emissivity is 0.3 will lead to a temperature that is about $100\,^\circ\mathrm{C}$ too low if the true emissivity is 0.1, for example with a clean aluminium surface, and a similar amount too high if the emissivity is really 0.9, for example with a heavily oxidised stainless steel surface.

Clearly, errors of this magnitude are generally unsatisfactory. If we make measurements instead at two wavelengths λ_1 and λ_2, and take the ratio of the measured signals I_1 and I_2, then

$$R = \frac{I_2}{I_1} = \frac{k_2\,\epsilon_2}{k_1\,\epsilon_1}\left(\frac{\lambda_2}{\lambda_1}\right)^{-5}\exp\left[-\frac{c_2}{T}\left(\frac{1}{\lambda_2}-\frac{1}{\lambda_1}\right)\right] \tag{7.2}$$

and the temperature obtained will be correct if the emissivity does not change from one wavelength to the other, that is, $\epsilon_1 = \epsilon_2$. The ratio of the emissivities may or may not be equal to unity, but the assumption appears at first sight rather more reasonable than that required for the single channel radiation thermometer. Instruments based on this principle are known as ratio or two-colour radiation thermometers. The variation of the signal ratio R behaves in a similar way to the output from a monochromatic instrument with an effective wavelength Λ, where

$$\frac{1}{\Lambda} = \frac{1}{\lambda_2} - \frac{1}{\lambda_1} \tag{7.3}$$

The price paid for this apparent advantage is a loss in sensitivity, as R varies more slowly with temperature than does the intensity in the single channel case, because the effective wavelength Λ must be longer than either λ_1 or λ_2. If we take a ratio thermometer with channel wavelengths at 600 nm and 750 nm, the effective wavelength is $3\,\mu\mathrm{m}$.

A glance at some plots of emissivity against wavelength will convince the reader that the emissivity is generally anything but constant, and the chance that ϵ_1 will be equal to ϵ_2 is remote. However, by increasing the number of channels to three, we will obtain the right answer for the surface temperature if the emissivity varies approximately linearly with wavelength. The principle can be generalised to the point that, if we make measurements in N channels with separate and distinct wavelengths, we should be able to find for some value of N a function with sufficient adjustable parameters to adequately describe the real behaviour of the spectral emissivity as a function of wavelength. We shall call the application of this assumption the 'exact fitting method', for reasons which will become clearer as we progress.

To analyse the validity of the assumption, we start by taking the logarithm of (7.1) and rearranging to give

$$-\lambda_i \ln\left(\frac{I_i \lambda_i^5}{k_i}\right) = \frac{c_2}{T} - \lambda_i \ln \epsilon_i \qquad (7.4)$$

The left-hand side of this equation, which we will designate y_i for convenience, that is,

$$y_i = -\lambda_i \ln\left(\frac{I_i \lambda_i^5}{k_i}\right) \qquad (7.5)$$

contains only known or measurable quantities. It will be considered that y_i is exact, that is, it is made up of the precise values for each parameter, and that the effects of uncertainties upon the calculated temperature may be investigated by adding a term Δy_i representing the overall uncertainty in y_i. The right-hand side of (7.4) contains the unknown properties of the surface, its emissivity and the temperature T. The second term is a function primarily of the wavelength, and varies only weakly with temperature. The first is clearly independent of the wavelength and a function of the temperature only. As the emissivity ϵ_i lies between zero and one, $\ln(\epsilon_i)$ is negative, and both sides of (7.4) are positive. So the left-hand side is represented by the y_i and contains known quantities, while the right-hand side contains the $(N+1)$ unknowns, that is, the temperature T and the set of spectral emissivities $\{\epsilon_i\}$. It will be seen that the y_i as defined have the dimensions of wavelength.

If we now consider a multi-wavelength radiation thermometer consisting of N channels, each of which can be represented by (7.4) at a distinct wavelength λ_i, then the set of equations contains $(N+1)$ unknowns and can only be solved if we can add one positive piece of information. The 'exact fitting' assumption discussed above provides this essential piece of information by suggesting that a function exists which relates emissivity and wavelength and which contains $(N-1)$ adjustable parameters.

Although the assumption does not need to be specific about the form of this function, in practice of course it must take a particular form for further progress to be made. All discussions of this problem have assumed that the logarithm of the emissivity may be represented by a polynomial in the wavelength

$$\ln \epsilon_\lambda = \sum_{j=0}^{j=M} \alpha_j \lambda^j \qquad (7.6)$$

as this both simplifies the analysis and corresponds to the relationships used generally for the two- and three-colour radiation thermometers mentioned above. Substituting into (7.4), and including the overall uncertainty Δy_i, we obtain

$$y_i + \Delta y_i = \sum_{j=0}^{j=M-2} \alpha_{j+1} \lambda_i^j - \frac{c_2}{T'} \tag{7.7}$$

where T' is the temperature obtained from the measurements. This will of course differ from the true temperature T as a result of the effects of the measurement uncertainties and of the systematic uncertainty introduced if the form or degree of (7.6) is incorrect.

To determine the coefficients a_j and T', we need an expression for the polynomial of degree $(N-1)$ through the N values of $(y_i + \Delta y_i)$. This is given by a Lagrangian polynomial, so we may write

$$\sum_{j=0}^{j=M-2} \alpha_{j+1} \lambda_i^j - \frac{c_2}{T} = \sum_{i=1}^{i=N} (y_i + \Delta y_i) \prod_{k=1, k \neq i}^{k=N} \frac{\lambda_k - \lambda}{\lambda_k - \lambda_i} \tag{7.8}$$

At this stage, we are only interested in assessing the uncertainty in the value of the temperature obtained and not at all in the coefficients a_j. It is only necessary, therefore, to consider the wavelength independent terms in this equation to evaluate T', that is,

$$-\frac{c_2}{T} = \sum_{i=1}^{i=N} (y_i + \Delta y_i) \Pi_i \tag{7.9}$$

where

$$\Pi_i = \prod_{k=1, k \neq i}^{k=N} \frac{\lambda_k}{\lambda_k - \lambda_i} \tag{7.10}$$

We first consider the uncertainty introduced by the assumption inherent in (7.6) in the absence of measurement uncertainties; that is, the measured values y_i are known exactly and the errors Δ_y are all zero. Substituting for y_i from (7.4) gives

$$-\frac{c_2}{T} = \sum_{i=1}^{i=N} \left(\lambda_i \ln \epsilon_i - \frac{c_2}{T} \right) \Pi_i \tag{7.11}$$

By considering the form of the Lagrangian polynomial which passes through N points of constant magnitude, it may easily be shown that

$$\sum_{i=1}^{i=N} \Pi_i = 1 \tag{7.12}$$

so that the constant term c_2/T may be removed from the summation on the right-hand side to give an expression for the error in the derived temperature T':

$$c_2 \left(\frac{1}{T} - \frac{1}{T'} \right) = \sum_{i=1}^{i=N} \lambda_i \Pi_i \ln \epsilon_i \tag{7.13}$$

This expression gives the well-known expressions for the simpler radiation thermometers. For a monochromatic device with N equal to one

$$c_2 \left(\frac{1}{T} - \frac{1}{T'} \right) = \lambda_i \ln \epsilon_i \tag{7.14}$$

For the two-colour device ($N = 2$)

$$c_2 \left(\frac{1}{T} - \frac{1}{T'} \right) = \frac{\lambda_1 \lambda_2}{\lambda_1 - \lambda_2} \ln \left(\frac{\epsilon_1}{\epsilon_2} \right)$$

$$= \Lambda \ln \left(\frac{\epsilon_1}{\epsilon_2} \right) \tag{7.15}$$

from (7.3).
For a three-colour device

$$c_2 \left(\frac{1}{T} - \frac{1}{T'} \right) = \lambda_1 \lambda_2 \lambda_3 \left[\frac{\ln \epsilon_1}{(\lambda_2 - \lambda_1)(\lambda_3 - \lambda_1)} - \frac{\ln \epsilon_2}{(\lambda_2 - \lambda_1)(\lambda_3 - \lambda_2)} \right.$$

$$\left. + \frac{\ln \epsilon_3}{(\lambda_3 - \lambda_1)(\lambda_3 - \lambda_2)} \right]$$

$$= \frac{\lambda_1 \lambda_2 \lambda_3}{(\lambda_2 - \lambda_1)(\lambda_3 \lambda_1)(\lambda_3 - \lambda_2)} \left[\lambda_1 \ln \left(\frac{\epsilon_2}{\epsilon_3} \right) + \lambda_2 \ln \left(\frac{\epsilon_3}{\epsilon_3} \right) \right.$$

$$\left. + \lambda_3 \ln \left(\frac{\epsilon_1}{\epsilon_2} \right) \right] \tag{7.16}$$

In each case the conditions for zero error may be obtained by inspection of these equations. For monochromatic and two-colour radiation thermometers, these are simple:

$$\epsilon_1 = 1 \qquad : \qquad \epsilon_1 = \epsilon_2 \tag{7.17}$$

respectively, as already stated above. For higher-order radiation thermometers, the condition depends upon the spacing of the channels. For a three-colour device with equal spacing, the condition reduces to

$$\epsilon_1 \epsilon_3 = \epsilon_2^2 \tag{7.18}$$

and this is equivalent to taking the ratio $I_1 I_3 / I_2^2$ of the channel signals. It may be worth emphasising at this point that, although these criteria are often discussed in terms of the emissivity dependence upon wavelength, the errors in multi-wavelength thermometry depend only upon the actual values of the spectral emissivities at the channel wavelengths. If, for example, the condition $\epsilon_1 = \epsilon_2$ is satisfied for a two-colour radiation thermometer at the wavelengths λ_1 and λ_2, the value and shape of the emissivity curve at any other wavelength is irrelevant. The value of (7.9) is that it enables errors arising from sets of values $\{\epsilon_1\}$ (including those between λ_1 and λ_2) to be calculated. By substituting sets corresponding to different analytical forms of emissivity dependence, it is possible to study the sensitivity of the temperature error to the validity of the fundamental assumption.

The term within the summation in (7.13) is related to the Nth divided difference of $\ln(\epsilon_i)$; this in turn is close to the Nth differential of the analytical form assumed for the logarithm of the spectral emissivity – in the present discussion a straightforward polynomial (equation (7.6)). If $\ln(\epsilon_i)$ is described exactly by a polynomial of degree $(N-1)$ or less, then the divided differences are zero and the calculated temperature T' is identical to the true temperature T.

We may note, however, that the coefficients $\lambda_i \Pi_i$ in (7.13) increase very rapidly as the number of channels increases. While the individual terms $\lambda_k/(\lambda_k - \lambda_i)$ within the product will probably fall within the range 2 to 5 for instruments with few channels, the separations $(\lambda_k - \lambda_i)$ will tend to decrease as N increases and the channels must be packed more closely. For quite moderate values of N, the coefficients within the summation become very large and, as they also alternate in sign, it is likely that quite small deviations from polynomial behaviour will produce significant errors in the calculated temperature.

In order to assess this possibility, various analytical forms can be adopted to generate sets of $\{\epsilon_i\}$, and the errors in T' for thermometers with various numbers of channels calculated. It is found that, although many functions give errors which reached a satisfactorily low level as N increases, there can be a large class of functions for which the errors actually increase with N and quickly become quite unacceptable. As there is no sound theoretical reason for selecting one function rather than another, and in many cases it is impossible on practical grounds to distinguish members of either class, it is clear that this represents a major drawback to the application of this technique.

If we consider the emissivity curve shown in Figure 7.1, this smooth monotonically decreasing curve is described by a very simple analytical function, and is reasonably representative of the variation of spectral emissivity of some metals at the wavelengths shown. The temperature errors introduced by the exact fitting assumption may be calculated using (7.13). The results for radiation thermometers with one to six channels are shown in Table 7.1 for a true surface temperature of 1500 K, with the channels spaced evenly through the wavelength range. It will be seen that the errors are initially moderate in size, but then begin to increase with the number of channels, becoming very large for a device with six channels. Moreover, the logarithm of the emissivity curve in Figure 7.1 can be approximated by a quadratic in the wavelength with a maximum deviation of about 0.1 % – a difference not discernable on the graph and very much less than the uncertainties in the best practical measurements. If the temperature errors are re-calculated with the

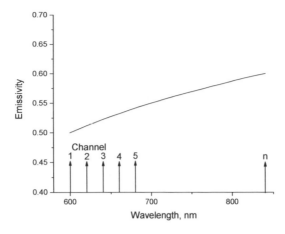

Figure 7.1: Emissivity curve for exact fitting example.

quadratic approximation in (7.13), it will be seen from the second column in Table 7.1 that they are of similar magnitude to the first case for radiation thermometers with up to three channels, but they become identically zero for greater values of N.

Table 7.1: Temperature error in kelvin for multi-wavelength radiation thermometers given a material with the spectral emissivity given in Figure 7.1 at 1500 K, for different numbers of channels

Number of channels	Systematic error from: Model curve	Quadratic fit	Measurement uncertainty
1	142	142	1.3
2	134	131	5.3
3	132	132	46
4	213	0	470
5	414	0	5180
6	884	0	59600

Note: The measurement uncertainty is for a 1 % uncertainty in photocurrent at each channel.

The significance of these results is that, in the absence of any sound experimental or theoretical knowledge of the correct form for the dependence of emissivity upon wavelength, it is impossible to estimate reliably the systematic uncertainty to be associated with the application of a multi-wavelength radiation thermometer of this type, even in the complete absence of experimental error. It should be stressed that this conclusion is not dependent upon the assumption of a polynomial form for the approximating function for $\ln \epsilon_i$. Other expressions with $(N-1)$ adjustable parameters could have been taken; the point is that there is no valid support for these either, and they would be equally prone to generate large systematic errors if they deviated from the real variation of emissivity with wavelength.

7.1.1 Effect of measurement errors

The fact that the application of the exact fitting assumption may lead to errors of unknown magnitude even in the absence of measurement error is in itself a major disadvantage. However, it is possible that a theory which gives the functional form of emissivity variation with wavelength may be developed in the future. Even if this should prove to be the case, there remains another fundamental limitation to the method. Let us assume that the polynomial assumption in (7.6) is valid, so that the systematic error from this source is zero, that is, from (7.13)

$$\sum_{i=1}^{i=N} \Pi_i \lambda_i \ln \epsilon_i = 0 \tag{7.19}$$

and consider the effects upon the measured temperature T' of the uncertainties Δy_i. The temperature uncertainty is now obtained by returning to the general error (7.8), substituting for y_i from (7.4) and (7.5), and applying the condition above, giving

$$c_2 \left(\frac{1}{T} - \frac{1}{T'} \right) = \sum_{i=1}^{i=N} \Pi_i \Delta y_i \tag{7.20}$$

where T' is the (incorrect) temperature affected by the various sources of measurement uncertainties present in the Δy_i. These may be obtained by differentiating (7.5). For example, for the uncertainty in the wavelength

$$\Delta y_i = - \left[\ln \left(\frac{\lambda_i^5 I_i}{k_i} \right) + 5 \right] \Delta \lambda_i \tag{7.21}$$

for that in the measurement of the signal

$$\Delta y_i = \lambda_i \frac{\Delta I_i}{I_i} \tag{7.22}$$

and from the calibration constants k_i

$$\Delta y_i = -\lambda_i \frac{\Delta k_i}{k_i} \tag{7.23}$$

In determining the overall uncertainty of a given radiation thermometer, the contributions arising in each channel from a variety of sources must be combined. The manner in which this can be effected depends largely on whether the components are random or systematic in nature, and whether they are independent or not. The analysis is not a simple one; the uncertainty $\Delta I_i / I_i$ in the measured signal in channel i is a combination of random errors from detector and photon noise, and the systematic errors from detector non-linearity, background noise (which may be subtracted out if measured separately), and scattering and diffraction effects. The similar term $\Delta k_i / k_i$ represents the drift in the channel calibration from detector ageing or variations in the ambient temperature. In a well-designed instrument, the calibration process should not add significantly to this. In addition, it may be necessary to consider the uncertainties in the nominal channel wavelengths. Although these are usually negligible for simple devices, the sensitivity to measurement errors of the more complex multi-wavelength radiation thermometers may mean that they need to be known to high accuracy.

To demonstrate the sensitivity of this type of multi-wavelength radiation thermometer to measurement errors, the temperature uncertainty produced by random independent uncertainties in the signals I_i has been calculated for instruments with one to six channels. The results have been included in Table 7.1, assuming a measurement uncertainty of 1 % and a true surface temperature of 1500 K. It will be seen that the overall uncertainty increases very rapidly with N, and only the simpler devices provide measurements of acceptable accuracy.

7.2   THE LEAST SQUARES APPROACH

The extreme errors produced by the forced multi-wavelength assumption in some cases is largely due to the over-fitting of the data points y_i inherent in the use of the Lagrangian polynomial – an approach which requires stringent checks and safeguards. An alternative assumption which seeks to avoid these problems is that, instead of fitting a polynomial of degree $(N - 2)$ exactly through the N values of y_i, polynomials of lower degree are least squares fitted, and that of lowest degree which adequately fits the data selected. The zero-order coefficient is again equated to the required temperature term c_2/T'. Alternatively, the channel intensities I_i may be fitted with a non-linear least squares technique to the Planck distribution, with the emissivity represented by a low-order polynomial in the wavelength.

While this revised procedure may well avoid the extreme errors of the forced fit, the magnitude of the resulting uncertainty is by no means evident. Again, this may be expected to arise for the same reason as before – the failure of the form of the model function, the low-order polynomial, to match the actual behaviour of the emissivity or $\ln \epsilon_i$ with wavelength, and the sensitivity of the calculated results to the generally small uncertainties Δy_i in the data.

The first misconception in the application of this method is that the form of the variation of emissivity can be deduced from a plot of the data y_i against wavelength. It is common to find statements in papers on this topic suggesting that it is possible to show from the data that, for example, the emissivity varied linearly with wavelength. The distinction between a plot of y_i and one of $\ln \epsilon_i$ is a rather subtle one, but it is nevertheless important. Consider the simulated data values y_i shown in Figure 7.2. These are intended to represent a set of measurements, with an uncertainty of about 1 %, on a nickel surface at 1250 K. To evaluate the temperature, we must determine the constant term in any function fitted to the data points. This is formally equivalent to extrapolating the function back to the y-axis. Extrapolation is a dangerous mathematical procedure for the unwary, and it is particularly inadvisable for radiation thermometry measurements, where the available wavelength range is relatively narrow, and the extrapolation range is of the same magnitude. The data points in Figure 7.2 appear to lie quite closely on a parabola, which corresponds from (7.4) to a linear variation of $\ln \epsilon_i$ with wavelength. This is shown as curve A in Figure 7.2, and the intercept on the y-axis gives an estimate for c_2/T. However, it will be seen that the data points are equally well fitted within the measurement region by curves B and C, which correspond to other simple functions for the emissivity, and that these give quite different intercepts on the y-axis. The temperatures obtained differ by about 80 K in this case. Other forms for the emissivity function, and larger experimental errors, may well give considerably greater values. Again, it is not possible from the experimental evidence to decide which of the curves A, B or C, or indeed any other, correctly represents the variation of emissivity with wavelength.

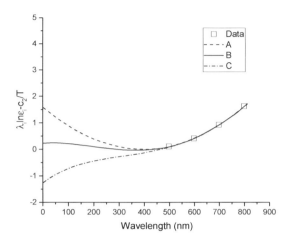

Figure 7.2: Extrapolation to determine c_2/T from the intercept of different functions with the y-axis. Functions that are indistinguishable over the fitted range can give substantially different values for T.

The effect of measurement uncertainties Δy_i upon the uncertainty in the derived temperature may be analysed rather easily in the least squares fitting case where the approximating series is a linear combination of functions. With a polynomial of degree M as the model function, the observation matrix $\boldsymbol{J}$ is of dimension $N \times (M+1)$, with elements

$$\boldsymbol{J}_{ik} = \lambda_i^{k-1} : i = 1...N \tag{7.24}$$
$$: k = 1...(M+1)$$

If the coefficients of the polynomial are represented by the vector $\boldsymbol{a}$, then the the $(M+1) \times (M+1)$ variance covariance matrix is given by

$$\boldsymbol{E}(\Delta\boldsymbol{a}\Delta\boldsymbol{a}^T) = (\boldsymbol{J}^T\boldsymbol{W}^2\boldsymbol{J})^{-1} \tag{7.25}$$

where $\boldsymbol{W}$ is an $N \times N$ diagonal matrix which weights the data points according to the estimated uncertainties Δy_i. For this

$$(\boldsymbol{W})_{ij} = 1/\Delta y_i \text{ if } j = i \tag{7.26}$$
$$= 0 \text{ otherwise}$$

The matrix $\boldsymbol{E}(\Delta\boldsymbol{a}\Delta\boldsymbol{a}^T)$ has the variance of the coefficients along the diagonal, with the covariances in the off-diagonal positions. Only the variance of the first diagonal element, which corresponds to the temperature term, is required here. This has been used to calculate the ratio F of the uncertainty Δy_0 in the intercept to that of the data points in the special case where the uncertainty Δy is the same for each. The uncertainty in the temperature itself is of course given by

$$\Delta y_0 = -c_2 \frac{\Delta T}{T^2} \tag{7.27}$$

Values for F for different numbers of points (N) evenly distributed over the range 500 nm to 1000 nm, and for various degrees M of polynomial, are given in Table 7.2. The first entry in each column corresponds to the exact fitting case. It will be seen that, as opposed to the suggestions of the proponents of least squares methods, the uncertainties are similar in the least squares case to those obtained with the exact fitting technique, and only fall away slowly as the number of points increases. It is evident from these figures that the approximating function must be restricted to a low-order polynomial in order to keep the effects of the inevitable measurement uncertainties within reasonable bounds. However, the likely effect of this limitation is that the difference between the polynomial approximation and the real emissivity function will be increased.

Table 7.2: *Errors in least squares method of multi-wavelength radiation thermometry*

Number of points	Ratio, F, for			
N	M=0	M=1	M=2	M=3
2	2.24			
3	2.20	10.40		
4	2.07	9.72	61.8	
5	1.95	9.29	52.4	400
6	1.84	8.92	48.5	311
7	1.74	8.58	46.1	276
8	1.66	8.27	44.3	257
9	1.58	7.98	42.8	244
10	1.52	7.72	41.4	238
.	.	.	.	.
.	.	.	.	.
50	0.73	4.01	22.5	128
100	0.52	2.90	16.4	94

The effects of systematic errors and corrections are also readily assessed. If they vary with wavelength, they may be represented by an approximating series

$$\frac{\Delta y}{y} = \sum_{j=0}^{j=\infty} b_j \lambda^j \tag{7.28}$$

so that the form of the polynomial fitted to the corrected data points would be

$$y - \Delta y = \left[\frac{c_2}{T} - \sum_{j=0}^{j=M} \alpha_j \lambda^{l+1} \right] \left[1 - \sum_{j=0}^{j=\infty} b_j \lambda^j \right] \tag{7.29}$$

The effect upon the intercept, at $\lambda = 0$, is therefore simply

$$y_0 = \frac{c_2}{T}(1 - b_0) \tag{7.30}$$

that is, only the wavelength-independent component of the systematic error, which acts as a constant offset to the data points, has an effect upon the temperature calculated.

7.2.1   Overall assessment of the technique

It may be concluded from the analysis above that only the simplest multi-wavelength radiation therometers are likely to be of any practical value, and even here any attempt to avoid an estimation of the actual emissivity characteristics of the surface will lead to uncertainties often too large to be acceptable. It should also be emphasised that only analytical forms for the variation of emissivity with wavelength that are smooth and simple have so far been considered. In practice, the emissivity may vary both rapidly and in a complex way with wavelength, as surface films and structure may be of the same dimensions as the wavelengths of the radiation being measured.

We may ask if any method which does not utilise additional information about the surface characteristics can do any better than the methods described so far. There is in fact a general argument which may be used to show that no approach is likely to be satisfactory without this additional information.

One of the faults in the multi-wavelength methods so far is that they place no limitations on any of the parameters to be determined. But in fact we know that no surface can have an emissivity greater than one. If we allow a margin for experimental uncertainties, calculations which indicate spectral emissivities significantly greater than unity must be incorrect. Similarly, even highly polished metal surfaces have emissivities greater than 0.01 to 0.02. We may therefore limit the solutions of the equations of multi-wavelength radiation thermometry to lie within these values, and this step in itself would remove the excessive errors found above.

If we have made measurements at a number of wavelengths and are fortunate enough to know, from an independent determination, the true temperature of the surface, we may calculate the spectral emissivities at these wavelengths. An example is shown in the points labelled 'T' in Figure 7.3; the solid curve is for clarity only and has no other significance. If our knowledge of the true temperature was in fact at fault, and we had been given incorrect temperatures T_U or T_L, for example, we would have calculated the other sets of results labelled in the figure. For moderate errors in the temperature, there is absolutely no way in which we can tell, from the distribution of the points or by any other means, that an error has occurred. Only when one or more of the calculated emissivity values approaches or exceeds the limits of unity or 0.01 to 0.02 given above might we suspect that something is wrong. Arguments based upon the shape of the curves through a given set of points are not valid. Although the measurement uncertainties are often low, between 1 % and 5 % for example, and the interpolating curves appear to be well defined, it must be realised that each point at an arbitrary temperature T' has been multiplied by a factor F_{ik} from its correct value, where

$$F_{ik} = \exp\left[\frac{c_2}{\lambda_i}\left(\frac{1}{T'} - \frac{1}{T}\right)\right]$$

so that if the emissivity appears to be constant or linearly varying with wavelength in one curve, it will not in another.

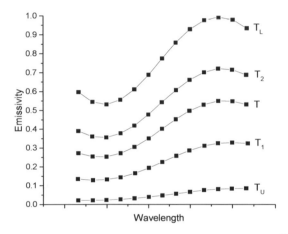

Figure 7.3: Permitted solutions to multi-wavelength measurements.

Another mathematical misconception which appears rather frequently is the idea that the minimum associated with a least squares fitting exercise must correspond with the correct values of the fitted parameters. This of course is not the case, and the minimum may depend upon such incidental factors as the validity of the model function, the weighting of the data points and the distribution of the measurement errors.

It is in fact possible to use this approach to obtain an estimate of the true temperature, with realistic estimates of the uncertainty in the result. If the lower limit T_L is given by the curve for which the largest spectral emissivity is equal to unity and the upper limit T_U by that for which the lowest emissivity is equal to the rather arbitrary value of 0.02, then we may be reasonably confident that the true value lies within the range T_L to T_U. In fact, if the spectral emissivity varies strongly with wavelength, this range may not be excessive. The method has a number of advantages, in that it is relatively insensitive to measurement errors, and it gives a positive estimate of the probable error. Once the wavelengths at which the maximum and minimum values of emissivity have been determined, there is no point in making measurements at other intermediate values, and these may be dropped. The wavelength range covered initially should of course be as wide as possible, to increase the chance of finding maximum and minimum values of the emissivity. However, in many applications the uncertainty in the result is simply too large, and this will be the case for all methods which do not utilise experimental values for one or more of the surface characteristics.

7.3 TWO-COLOUR RADIATION THERMOMETRY

It has been noted earlier in this chapter that the technique of two-colour or ratio thermometry does not in general give rise to the excessive measurement and systematic errors characteristic of those methods employing a greater number of channels. While it is not as potentially sensitive or accurate as monochromatic radiation thermometry, there are applications in which it has distinct advantages. One example is provided in the manufacture and treatment of aluminium, whose spectral emissivity is not only rather low but very dependent upon the roughness and degree of oxidation of the surface and consequently varies with time. In addition, this technique enables estimates of the temperature of small objects, i.e., those which do not fill the field of view of available thermometers, to be made, as the value obtained depends upon the relative spectral distribution rather than the absolute intensity in each channel. For the same reason, a correction is obtained for the effects of absorption or scattering in the optical path, from smoke, windows or absorbing gases, as long as this is the same in each channel.

From the Wien approximation, which is generally valid for the uncertainty levels and temperature ranges appropriate to two-colour radiation thermometry, we may write for the signal ratio R

$$\ln R = \ln \frac{I_2}{I_1}$$
$$= \ln \frac{k_2}{k_1} + \ln \frac{\epsilon_2}{\epsilon_1} - 5 \ln \frac{\lambda_2}{\lambda_1} + \frac{c_2}{T} \left(\frac{1}{\lambda_1} - \frac{1}{\lambda_2} \right) \quad (7.31)$$

The effective wavelengths λ_1 and λ_2 allow for the fact that the bandwidth of radiation detected in each channel may not be negligible, and may be calculated in a similar way to the monochromatic case. The relationship between the uncertainty ΔR in the measured ratio and that produced in the calculated temperature is easily obtained by differentiating the expression

$$\frac{\Delta R}{R} = \frac{c_2}{T} \left(\frac{1}{\lambda_2} - \frac{1}{\lambda_1} \right) \frac{\Delta T}{T} \quad (7.32)$$

This is similar to that for the monochromatic radiation thermometer, but with a much longer overall effective wavelength Λ as defined in (7.3). If we define λ_0 as the geometrical mean of the channel wavelengths, (7.32) may be re-written in the form

$$\frac{\Delta R}{R} = \frac{c_2}{\lambda_0 T} \left(\frac{\lambda_1 - \lambda_2}{\lambda_0}\right) \frac{\Delta T}{T} \tag{7.33}$$

Compared to a monochromatic thermometer, therefore, the sensitivity is reduced by a factor equal to the ratio of the channel separation $\lambda_2 - \lambda_1$ to the mean wavelength λ_0. The main result of this in practice is that the temperature uncertainty is sensitive to errors or changes arising within the instrument, as well as those, discussed earlier, from the ratio of the emissivities. To minimise this, it is necessary in the design of a two-colour radiation thermometer to consider very carefully the effects of ambient temperature and ageing upon the characteristics of the critical components, and to ensure that the mechanical mounting of these is particularly rigid. In addition, a two-colour device should be checked and calibrated at more frequent intervals than necessary for a monochromatic instrument.

The detailed dependence of the temperature uncertainty upon the measurement and wavelength uncertainties in each channel may be derived by differentiating (7.2),

$$\frac{\Delta T}{T} = \left(\frac{c_2}{\Lambda T}\right)^{-1} \left(\frac{\Delta I_2}{I_2} - \frac{\Delta I_1}{I_1}\right) \tag{7.34}$$

It will be seen that it is possible for errors to cancel if they affect each channel equally. For example, contamination of the optical surfaces by a deposit whose transmission was wavelength independent would not affect the reading. Unfortunately, complete cancellation of errors occurs only too rarely in practice. In some cases, the uncertainty may be of different sign, and the components reinforce each other.

We now consider the choice of operating wavelengths for a particular application. The first priority is to find whether the emissivity or the ratio ϵ_2/ϵ_1 is known in any region of the spectrum. If it is, then this is a very strong point in favour of selecting these wavelengths. However, because of the lower sensitivity of two-colour radiation thermometers, as indicated in (6.25), the range of acceptable wavelengths for a given source temperature is smaller than that for monochromatic devices. It is in general undesirable to operate at less than proportional sensitivity, i.e., with $\Delta T/T - \Delta R/R$. If we take a typical fractional bandwidth of 0.2, this corresponds to a mean wavelength λ_0 given by

$$\frac{c_2}{\lambda_0 T} \approx 5 \tag{7.35}$$

which is close to the wavelength of the maximum in the Planck distribution. It is sometimes stated that the channel wavelengths should not straddle the peak, as all sensitivity will then be lost. This is not the case, as the sensitivity does not fall to zero until very much longer wavelengths, and if the fractional bandwidth is made large, operation on the long wavelength side of the peak is quite possible.

Nevertheless, for a given emissivity ratio and fractional bandwidth, the temperature uncertainty is minimised by working at the shortest wavelengths at which adequate signal intensity is available.

In fact, many designs of two-colour radiation thermometers have poor sensitivity as a result of adopting too small a value for the fractional bandwidth. There are indeed technical problems with increasing the channel separation to the point where the wavelengths are in distinctly different regions of the spectrum, and it then becomes difficult to control the contributions from wavelength-dependent sources of error. But increasing the fractional bandwidth is a much more sensible step than minimising it in order to keep the ratio ϵ_2/ϵ_1 close to unity. From (7.15), the temperature uncertainty arising from the unequal values of the emissivity at the channel wavelengths may be written as

$$c_2 \frac{\Delta T}{T^2} = \Lambda \ln \frac{\epsilon_2}{\epsilon_1} \tag{7.36}$$

so that if the logarithm of the emissivity is a linear function of wavelength, that is

$$\ln \epsilon_i = a + b\lambda_i \tag{7.37}$$

then the temperature uncertainty is

$$c_2 \frac{\Delta T}{T^2} = b\lambda_1 \lambda_2 \tag{7.38}$$

that is, it is independent of the channel separation. In practice, of course, the error may be greater or less than this, depending on the precise variation of emissivity with wavelength.

The low temperature limit for a two-colour radiation thermometer is determined by the poorer of the signal-to-noise ratios in the two channels, generally that at the shorter wavelength. This may be improved for selected channel wavelengths and photo-detectors, by increasing the throughput of the thermometer, either by using large aperture optical components or by increasing the optical bandwidth of one or both channels. The first tends to be expensive in terms of component cost, and limited in advantage if the thermometer was well designed in the first place. Broadband optical filters are, on the whole, not much more expensive than narrow-band filters. The bandwidth may be increased to the point where the spectral response curves for the two channels overlap moderately. Beyond this point, there is a law of diminishing returns. While the additional radiation improves the signal-to-noise level, the sensitivity of the thermometer is reduced because each channel is detecting radiation from the same spectral region.

7.3.1 Sources of uncertainty

7.3.1.1 From the variations of emissivity

The uncertainties arising from the failure of the fundamental assumption of two-colour radiation thermometry, that is, $\epsilon_1 = \epsilon_2$, may be minimised in those cases where some information is available in the literature, either by selecting channel wavelengths whose emissivities are nearly equal, or by making a correction of the form required by (7.15). It is necessary to issue a word of warning about the accuracy of this procedure; it is frequently found that the published values are insufficiently accurate or that they do not adequately represent the surface under examination.

In those cases where no reliable information is available, it may be possible to obtain an estimate for the emissivity ratio if the temperature of the surface can be determined by another means, for example with a thermocouple or resistance thermometer. On the whole, the emissivities and emissivity ratios of stable surfaces do not change rapidly with temperature, and a single measurement may be applied over a considerable temperature range if the surface is stable, that is, no physical or chemical changes occur. If no estimates or measurements of the emissivity ratio are possible, the magnitude of the resulting uncertainty cannot be estimated, and the errors may be very high indeed. If a two-colour thermometer with channel wavelengths of 500 nm and 600 nm were used to measure the temperature of a clean gold surface close to its melting point (1064 °C), an error of about 200 °C would be found. This arises because the emissivity of gold drops very rapidly, from 0.5 at 500 nm to 0.18 at 600 nm. In this case the colour of the surface might (on a good day!) alert the user of the instrument. No such visual clues are available of course for devices operating at infrared wavelengths.

Temperature uncertainties may be introduced in two-colour radiation thermometers by any change in the operating conditions which is wavelength dependent or differs for the two channels. Common examples are given by wavelength-dependent scattering from particles – smoke, fog, haze and so on – in the optical path, by absorption by water vapour or carbon dioxide in the atmosphere, and by changes in the transmission of optical components from contamination.

7.3.1.2 Other characteristics of the source

Because two-colour radiation thermometry effectively is a measure of the spectral distribution from the source, the temperature obtained will be affected by the presence of temperature gradients over the target area. In this case, the measured value tends towards the highest temperature present. It is unaffected by relatively cold areas, as these do not contribute to either channel. For this reason, it is possible in principle to measure the temperature of objects which occupy only a small fraction of the field of view of the thermometer. In practice, it is advisable to ensure that the object occupies more than one-third of the target area, especially for temperatures towards the lower end of the operating range. This reduces the possibility of scattered light from other sources entering the instrument, and prevents the photo-detectors from working at very low light levels where photon and background noise may be a problem.

To illustrate the magnitude of the effects produced by temperature gradients, Figure 7.4 shows the calculated readings to be expected from a monochromatic radiation thermometer with an effective wavelength of 600 nm and a two-colour thermometer with channel wavelengths of 600 nm and 800 nm. In both cases the circular target area of radius R is filled by an object whose temperature is constant at 1500 K within a smaller central circle of radius r, and then falls away linearly to room temperature at the edge of the target area. The expected reading is plotted as a function of the ratio r/R. It will be seen that the two-colour thermometer is much less sensitive to the presence of gradients than the monochromatic device. When the area at constant temperature is only 10 % of the total ($r/R = 0.32$), the error with the two-colour instrument is only 25 K, whereas the monochromatic thermometer reading is in error by about 170 K.

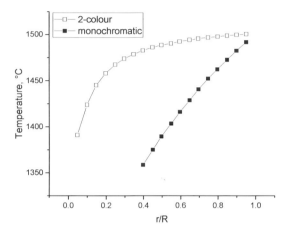

Figure 7.4: Error due to target non-uniformity.

While the two-colour radiation thermometer is relatively robust in this aspect of its operation, it is important that each channel should view the same area of the target. In the presence of temperature gradients, quite small differences in the area viewed can produce significant errors. This is not so important when the thermometer is looking at a small object not filling the field of view, but in this case there may be errors arising from the fact that some of the radiation is emitted from areas at high angles to the optic axis of the instrument. Not only will this radiation be polarised, which may cause an error if the thermometer is sensitive to the state of polarisation of the detected radiation, but the angular dependence of the radiance is wavelength sensitive.

7.3.1.3   The measurement accuracy and stability

The main factors which affect the accuracy of measurement have already been covered; in addition, those given for the monochromatic thermometers are equally relevant here. Since they are often neglected, we would draw attention to the effects of scattered light and reflections inside the instrument, which can produce unsuspected systematic errors. Because of the sensitivity of the measurement to any changes in the relative response of the two channels, it is particularly important to check any factor which may cause this. The first is produced by changes in the ambient temperature. The relative response of silicon photodiodes varies with temperature, and other errors can be produced by the variation in transmission and effective wavelength of colour filters with temperature. Although a correction may be applied using a measured ambient temperature coefficient, this is only valid when the radiation thermometer is in equilibrium, and the presence of internal temperature gradients may give additional errors, particularly for those designs with two photodetectors, and with the channels physically separate. The second factor concerns the ageing rates of the components, i.e., it depends upon the change in characteristics with time rather than, as just discussed, with temperature. In general, components age steadily with time, gradually becoming more stable throughout their operating life, but this is not always the case. We may include in this category the variations of readings that may occur with the state of the batteries in portable instruments, and note that a step in the calibration curve may follow battery replacement or re-charging. For these reasons, it is especially important to re-calibrate this type of thermometer at frequent intervals, or, at the very least, to check the stability of its calibration at one or more points on each range against a source whose characteristics are known to be stable.

7.4   OTHER METHODS OF MULTI-WAVELENGTH RADIATION THERMOMETRY

In this section we shall consider a rather different approach to multi-wavelength radiation thermometry than those described earlier in this chapter. Instead of seeking general designs of instruments which attempt to determine the true temperature of surfaces whose emissivity is not known, we look at ways in which established values may be used to calculate the correct temperature from the observed readings. If the emissivity at a given wavelength, or the emissivity ratio at a pair of wavelengths, is constant for the surface in question, there is of course no problem, as the techniques of monochromatic and two-colour radiation thermometry may be applied and appropriate corrections made to the results. If this is not the case, other possibilities are available.

Let us consider, for example, the effect of multiplying the channel equation (7.4) by an arbitrary constant a_i, giving

$$a_i y_i = a_i \ln \epsilon_i - a_i \frac{c_2}{\lambda_i T} \tag{7.39}$$

These equations may be solved in a number of ways to give the true temperature T. Being aware of the problems involved with several channels, we will restrict ourselves to the two-channel case, when we may obtain

$$\frac{c_2}{T} \left(\frac{a_2}{\lambda_2} - \frac{a_1}{\lambda_1} \right) = a_2 \ln \epsilon_2 - a_1 \ln \epsilon_1 - a_2 y_2 + a_1 y_1 \tag{7.40}$$

Only the ratio $a = a_2/a_1$ is significant, and so this becomes

$$\frac{c_2}{T} \left(\frac{a}{\lambda_2} - \frac{1}{\lambda_1} \right) = \ln \frac{\epsilon_2^a}{\epsilon_1} - a y_2 + y_1 \tag{7.41}$$

which gives the true temperature if $\epsilon_2^a = \epsilon_1$. The value of the parameter a may be chosen to suit the measured variation of the emissivity in each channel, even if this changes with the state of the surface. In two particular cases, the instrument behaves as a monochromatic device ($a = 0$), or as a conventional two-colour device ($a = 1$), but it is not restricted to lie within this range.

It is not necessary to restrict the correction to analytical relationships of this type, however. Any empirical function found experimentally between the emissivities in the two channels may be pressed into service if it has been established that it will give meaningful results. The basic requirement is that if we convert the signals in each channel to radiance temperatures, the difference between the two, usually in the form $(1/T_{R1} - 1/T_{R2})$, may be used to generate a correction to $1/T_{R1}$. It is, though, necessary for the difference to be a monotonic function of the surface emissivity. The correction should also give the correct answer for black-bodies, as these are normally employed for calibration.

CHAPTER 8

Emissivity correction methods

8.1   INTRODUCTION

As seen in the earlier chapters, one of the main problems in the application of radiation thermometry is the fact that very few real surfaces behave as black-bodies. In general their emissivities are much less than unity, and a sizeable correction needs to be made to the measured spectral radiance and hence to the indicated temperature. Moreover, in this situation it is possible for thermal radiation from other hot sources in the vicinity to be reflected from the target area into the thermometer, producing a second error. In this chapter we shall describe the techniques which may be employed to reduce these uncertainties. The topics of ratio and multiwavelength radiation thermometry which are also claimed to attack the emissivity problem have been covered in Chapter 7; here we shall deal mainly with single band radiation thermometers and the use of any additional information which may be ascertained for the system under study.

It will be assumed that the surfaces to be investigated are reasonably flat and not macroscopically rough, as this class of surface presents an extra dimension of complexity as discussed briefly in Chapter 3. For example, the degree of polarisation of emitted and reflected radiation from a flat surface is a function of the angle of observation. It is often possible to minimise polarisation effects by making measurements along a direction close to the normal to the surface. For macroscopically rough surfaces, some area of the surface will always be at a large angle to the direction of observation, and polarisation effects may well be important. In some cases, as we shall see, it is possible to utilise polarisation effects to derive useful information about the surface characteristics.

The first and by far the most important point to make is that there are no universal solutions to the problem of emissivity compensation. The situation is quite different from that of contact thermometry, where a few simple rules are often sufficient to select the type of thermometer required for a given application, and to ensure that the results obtained reach the required level of accuracy. The optical characteristics of real surfaces have been discussed in Chapter 3. In a particular radiation thermometry application, it may ideally be necessary to know not only the emissivity of the surface at the operating wavelength, but also the angular distribution of the reflected radiation and the degree of polarisation of both the emitted and the reflected radiation. This depth of practical information is rarely available, and the various techniques which have been developed for emissivity correction attempt to simplify the situation by making appropriate and specific assumptions about the behaviour of the surface and its environment. The first step in the choice of a technique is therefore to establish which set of assumptions best match the application in question.

It is to be stressed that preliminary experiments are of major value in this respect. Even very simple tests may yield useful information, for example the degree to which the surface reflectance is diffuse or specular in nature, and whether the process produces changes in the structure of the surface, and hence its emissivity. Even when an appropriate technique has been identified, it must not be forgotten that the inherent assumptions cannot represent the real system exactly. Practical measurements with the selected radiation thermometer *in situ* are always to be recommended, to indicate the magnitude of the uncertainties introduced by this failure. The presence of systematic errors, for example, is often indicated by a lack of measurement reproducibility. An independent measure of the surface temperature with an alternative thermometer which may be inappropriate for normal working practice enables errors to be quantitatively assessed.

In considering the required accuracy, it is fortunate that in most applications the lowest uncertainties are only essential over a relatively narrow temperature range. Outside this, for example in pre-heating treatment or when allowing the plant to cool down, a rough indication of temperature may be quite adequate. Each radiation thermometer model operates over a limited range and tends to achieve its best performance towards the middle of this range. Clearly it is possible in selecting a thermometer to match its specification to the required performance in this respect. In some manufacturing processes, it is necessary only to return the system to some pre-determined condition, and no absolute temperature measurement is needed, that is, reproducibility is more important than absolute accuracy. It is not then necessary to know or correct for the emissivity of the surface, as long as it is stable at the maximum temperature of the process. However, this approach should only be adopted with caution, as subsequent changes in the materials involved, the replacement or re-calibration of the thermometer, or the requirement to establish further installations may introduce systematic offsets which would have been avoided by an absolute temperature measurement.

Most of the information available in the scientific and technical literature deals with measurements on metal surfaces, reflecting not only their commercial importance but the variety of associated problems. At low temperatures, they tend to have shiny surfaces with low emissivities. The correction for emissivity is therefore large, and specular reflections from extraneous sources can also give rise to very large errors. At higher temperatures, the effective emissivity may increase, either from physical processes such as crystal grain growth which roughens the surface, or from chemical reactions such as oxidation, which can produce a high emissivity layer on the metal surface. When the thickness of the layer is of the same order as the operating wavelength of the radiation thermometer, the surface characteristics may vary dramatically. The main advantages of experiments on metal surfaces are that they are quite opaque to radiation at all wavelengths, and that the thermal conductivity is high so that temperature gradients are not normally a major problem.

8.2   CORRECTION WITH EMISSIVITY VALUES

If the surface acts as a black-body, there are no problems of the nature which concern us here. The emissivity correction is zero, and radiation from any other source is absorbed at the surface and cannot be reflected into the radiation thermometer. The situation is almost as satisfactory if the surface characteristics are stable and their values are well known, as accurate corrections for the emissivity and the contribution from reflected radiation may usually be determined without difficulty. This is more likely to be the case at low temperatures where chemical and physical changes in the surface are less likely to occur, and at long wavelengths where those changes have a smaller effect upon the emissivity.

The maximum uncertainty permitted in the emissivity may of course be calculated readily from the temperature uncertainty required. The equations which relate the two have been derived in earlier chapters for the different designs of radiation thermometer in common usage. It is sufficient in general to adopt a simple model for the behaviour of each type. Even broadband radiation thermometers can usually be treated as narrow-band devices operating at their mean or effective wavelength. To minimise the effects of uncertainties in the surface characteristics, the radiation thermometer should operate at as short a wavelength as possible, consistent with the need to obtain adequate signal-to-noise ratios. In some cases, this general rule may be overridden by the behaviour of the surface under study, which may have a high or a stable and well-characterised emissivity within specific wavelength bands.

However, when the scientific literature is searched for emissivity values for appropriate conditions of temperature, wavelength and polarisation state, the paucity of available data very rapidly becomes apparent. For many materials the researcher is fortunate to find a single value at one wavelength. Even for those materials where several investigations have been made, usually metallic elements or simple compounds, it is clear that the reproducibility of the results from one sample to the next is often poor and depends critically upon the preparation of the surface, or its thermal history. This is especially the case, as might be expected, at high temperatures and for wavelengths in the visible. A not atypical set of values is shown for molybdenum in Figure 8.1, taken from the thermophysical properties 'bible' by Touloukian and DeWitt [14]. It will be seen that there are substantial variations between samples prepared with different techniques, even though molybdenum itself may be regarded as a well-behaved and thermally stable material. It is possible that

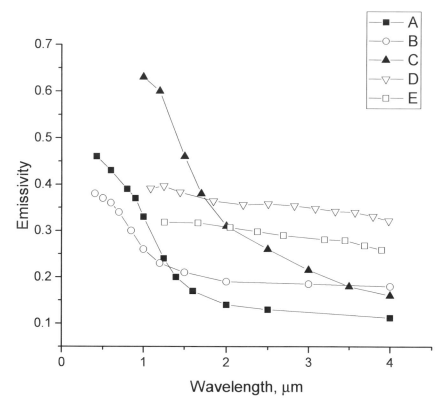

Figure 8.1: Normal spectral emissivity for various molybdenum samples.
A – from powder, hot pressed, sintered and polished, measured in argon at 1600K
B – as A, measured at 2800 K
C – cleaned foil, heated for eight hours, measured in argon at 1244 K
D – commercial sample, polished and washed, measured in argon at 1386 K
E – as D, some surface oxidation, measured at 1284 K.

the spread of values is partly due to systematic errors present in the measurement techniques employed, but it is worth noting that often trends can be identified for samples prepared in a particular way that may guide the user to select particular curves.

Despite the enormous amount of data collated by Touloukian and deWitt there are actually a very limited number of materials for which a reasonably complete range of values are available, and it is necessary to screen the data to obtain emissivities relating to a surface with a consistent structure. Typically uncertainties in the emissivities are in the range 5 % to 20 %. At 1000 °C, the equivalent temperature uncertainties for a narrow-band radiation thermometer with an operating wavelength of 650 nm is around 5 °C to 10 °C. At longer wavelengths the uncertainties increase proportionately. Battuello and Ricolfi [21] studied different radiation thermometer types for materials with reliable emissivity data and while the best performance was obtained with a short wavelength ratio thermometer, this does not imply that this class of instrument is necessarily the best choice for other materials, as the fluctuations in the uncertainty are likely to be greater than in the narrow-band case. In addition, short wavelength radiation thermometers are only suitable for use with sources at relatively high temperatures, where surface changes are more likely to occur. In one or two cases, tungsten for example, the surface properties were considered sufficiently stable and well established that acceptable uncertainties could be claimed even for broadband and total radiation thermometers.

For other materials of interest, a limited number of results may be available, but at wavelengths and temperatures other than those required. It may be possible to adapt these to the application in hand. First, it should be noted that the spectral emissivity changes to a greater extent and less predictably with wavelength as opposed to temperature, and it is dangerous to extrapolate values from one wavelength range to a significantly different one. This is particularly the case at relatively short wavelengths, in the visible and near infrared. At wavelengths beyond 10 μm many surfaces, especially metals and alloys, behave classically and the emissivity decreases slowly and smoothly with wavelength. However, it is then rather low and the overall temperature uncertainty to be expected from the extrapolated value is poor. The extension of emissivity values to other temperatures is, on the other hand, a more reliable process if conducted with care. Typical values of the emissivity temperature coefficients are less than $10^{-4}\,\mathrm{K}^{-1}$, so that known emissivities may be corrected over a range of a few hundred degrees without introducing a major error as long as the surface is not greatly changed by decomposition or oxidation.

If a survey of the technical literature fails to reveal any useable information, then the next step is to decide whether the surface characteristics are stable at the temperatures encountered during processing. In some cases it is all too obvious that significant changes take place, and it is unlikely that the emissivity will remain constant. If the process temperature is relatively low, and the appearance of the surface does not vary, then it may be worthwhile undertaking practical tests to determine the emissivity and the extent to which it may change during processing. While the emissivity may be obtained in independent experiments, there are distinct advantages in employing the radiation thermometer and associated equipment *in situ* for this purpose. Considerable information about the behaviour of the material and the equipment may be obtained in this way. At the very least, systematic errors in the measurement may be partially cancelled in this manner.

The simplest technique for estimating the emissivity is to place a calibrated contact thermometer or a small black-body in thermal contact with the surface, and then adjust the emissivity control on the radiation thermometer until the temperature reading from the surface agrees with that of the thermometer, or is the same as that from the black-body with the emissivity control set to unity. The main practical problem is to ensure that the thermal contact between the calibrating device and the surface is good, and that the temperature difference between them is small. With materials of high thermal conductivity this is not usually too difficult, but with those of low conductivity it may be necessary to halt the process to allow the system to reach thermal equilibrium. Again the reproducibility of the results under varying conditions provides an acid test of their validity. Although it may be sufficient to measure the emissivity at a single temperature to effectively calibrate the thermometer system, a little extra effort will yield valuable information on the stability of the surface, the uncertainty of the measurements and the variation of emissivity with temperature.

8.3   SURFACE MODIFICATION

In many cases, if not the majority, not only will little useful information be discovered in the literature, but it is clear that the surface characteristics change during the process to be monitored or controlled. It may be found from tests of the kind recommended above that a pattern in the emissivity changes can be observed. The reproducibility of these, however, often depends quite critically upon the procedures employed, and it can be dangerous to assume that a pattern established within a limited set of tests can be extended to other environments. The tests do of course give some indication of the range of variations likely to be encountered.

Changes in the emissivity of the surface may be produced by a wide variety of effects. These include chemical changes, such as oxidation and thermal decomposition (pyrolysis), and physical modifications, for example the increase in roughness produced by grain growth in metals. In some cases the two may be associated; the different characteristics of chemical products may induce cracking or fragmentation of the surface, increasing its roughness and hence its effective emissivity. In other cases, thermal cycling of fragile coatings can reduce emissivity as the surface degrades. The thickness of a surface layer is important in comparison to the wavelength of the measuring device. In general, a layer of thickness much less than the wavelength will not affect the optical properties, that is, it will effectively be transparent. The contribution from a layer a few wavelengths thick will depend upon the absorption coefficient, and may vary from transparency to almost complete absorption. As the result may vary rapidly with the wavelength, the effect upon the emissivity becomes very difficult to predict.

One simple way of overcoming these problems is to modify the surface in such a way that its properties become predictable and stable. Ideally, it should behave as a black-body so that undesirable signals from reflected radiation also disappear. Small holes may be drilled into the surface to act as black-body cavities. These should have a diameter rather larger than the target size of the radiation thermometer, and be at least four times as deep. The effective emissivity is increased if the sides of the hole are rough – threaded holes in mechanical components are very suitable for this purpose. If the thermal conductivity of the target material is low, large temperature gradients normal to the surface may exist, and the reading obtained by this means will correspond more closely with the bulk temperature of the object, rather than that of the surface. This may or may not be desirable.

It may of course be inconvenient to drill suitable holes. However, many objects have as a normal part of their structure re-entrant cavities or even simple corners which may be used to provide an increase in the effective emissivity compared to a section of flat surface. Macroscopically rough surfaces usually possess deep cracks or fissures which act as high emissivity sources. If the object is moving, or the radiation thermometer is scanned across the surface, the maximum signal is obtained from these areas. Some radiation thermometers have modified detection systems called 'peak pickers' which lock on to the signal maxima to make use of this effect. Of course the user should be aware of the effects of temperature gradients upon the validity of the results. In the case of thick layers of oxide and slag on metal surfaces, for example, not only may the uppermost layer be very much cooler than the metal, but large cracks may reveal unoxidized metal, which can be hot but has a much lower emissivity than the adjacent slag.

In the case of liquids or powdered solids, it may be possible to generate cavities with moderate emissivities by controlling the flow of the material. The temperature of molten steel was at one time measured with a 'blow-tube', through which a stream of compressed nitrogen was blown to produce a cavity in the molten liquid. A radiation thermometer at the cooler end of the tube collected radiation from the cavity, which had an effective emissivity of about 0.7-0.8. The low value was partly due to the low emissivity of the cavity walls and partly due to the cooling effects of the gas stream.

If it is not possible to find or form cavities which will give approximate black-body behaviour, the surface itself may be modified to give an emissivity which remains stable under the conditions of the process. A high value is of course to be preferred. It is only necessary to treat a small area, a little larger than the target area of the radiation thermometer, to give some latitude for any problems in aiming or focusing. The modification may be effected by forming or adding a suitable layer to the surface which absorbs strongly at the operating wavelength. It should not be too thick, however, and its thermal conductivity must be reasonably good, or a significant temperature difference may exist between the emitting surface and the underlying material. On the other hand, if the thermal conductivity of the bulk material is itself low, the presence of a local area of high emissivity on the surface with a higher radiant energy loss will distort the temperature distribution in its vicinity.

Figure 8.2 shows the equilibrium temperature gradients produced on the upper surface of a thick sheet by a circular target area of 16 mm radius. The emissivity of the target was taken as unity, while that of the uncoated material was 0.3. The lower surface of the sheet was assumed to be held at a fixed temperature of 1500 K. Only radiative heat loss was included in the calculation, that is, it was assumed that convective heat loss was negligible. The radial temperature gradients are of similar form for each value of the thermal conductivity K, although the magnitude of the temperature drop across the sheet varies roughly inversely with K. These curves represent the true temperature of the surface; if the radiance temperature were to be measured with a radiation thermometer, the values shown in the target area would be unchanged but those outside would be much lower, because of the lower emissivity.

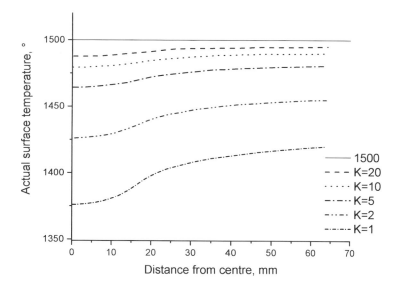

Figure 8.2: Radial temperature gradients produced by the addition of a 16 mm-radius circular area of high emissivity to the surface of materials with finite thermal conductivities K.

While most metals and high density oxides such as alumina, for example, have thermal conductivities in the 10s and 100s $W\,m^{-1}\,K^{-1}$, brick and glass have values around unity, and expanded materials such as foams may have very low conductivities indeed. The presence of materials with poor thermal conductivities may give rise to major problems in thermometry in general, not only because large temperature gradients may be produced but also because the associated thermal time constants may become very long and the system may indeed never reach a true equilibrium state. Both effects tend to obscure the significance and utility of the results obtained. For example, the temperature drop across a thin layer of scale on the surface of a steel ingot may amount to several tens of degrees. Although a correction may be calculated if the thickness of the layer and the external conditions are estimated, the uncertainty involved may still result in increased costs in keeping the ingot up to temperature for subsequent treatment.

The surface treatment required to produce a suitable target area may take one of two forms. If the material is a clean shiny metal which shows large changes in the emissivity as it oxidises, a small area may be prepared either by heating with a suitable torch or with an oxidising chemical so that the emissivity of that area remains relatively stable throughout the ensuing process. At lower process temperatures it is often simpler and more effective to apply a surface coating. Black coatings are particularly appropriate. However, the overriding requirements are that the emissivity should be high at the radiation thermometer wavelength and stable during the process concerned, and many other paints and coatings may be investigated for suitability. In some applications other requirements may have to be met. For example, it may be necessary that no chemical reactions should take place between the coating and the bulk material, and that the coating should be easily removed at the end of the process. Although the surface modification techniques are simple, effective and economical in concept and practice, they are difficult to apply to moving or inaccessible surfaces. Even then, it may be possible to add a thin stripe or spots of high emissivity for temperature measurement or control.

8.4   REFLECTION OF RADIATION FROM THE SURFACE

There are other techniques which may be used to increase the effective emissivity of a surface. Consider the effect of placing a cold mirror of reflectance ρ_M above a flat surface at a uniform temperature T which is viewed at a small angle θ from the normal by a radiation thermometer (Figure 8.3). The signal obtained from the thermal radiation emitted directly from the target of area A is

$$I_0 = k\epsilon_\lambda L_\lambda(\lambda, T) A \cos\theta \omega_p \tag{8.1}$$

If we assume for simplicity that the acceptance angle ω_p of the thermometer is small, and that the hot surface acts as a specular reflector, then the effect of the mirror generates a number of additional reflections which allow areas B and C to make contributions I_N to the total signal, where

$$I_N = I_0[\rho_M(1 - \epsilon_\lambda)]^N \tag{8.2}$$

The term within the brackets represents the reflection losses at the mirror and surface for the radiation originating at these secondary sources. These become more out of focus and larger in area as the number of reflections increases. For a large number of reflections the signal tends to a limiting value I_T.

$$I_T = I_o \sum_{N=0}^{N=\infty} [\rho_M(1 - \epsilon_\lambda)]^N$$

$$= \frac{I_0}{1 - \rho_M(1 - \epsilon_\lambda)}$$

$$= \frac{\epsilon_\lambda}{1 - \rho_M(1 - \epsilon_\lambda)} kL_\lambda(\lambda, T) A \cos\theta\omega_p \tag{8.3}$$

In the extreme case where the number of reflections is large and the mirror reflectance is unity, the surface emissivity ϵ_λ disappears from the right-hand side of this equation, and the signal corresponds to that from a black-body at the temperature T. These conditions are extremely unlikely to be obtained in practice, as the mirror can only be of finite size and must be placed at a reasonable distance from the hot surface to avoid contamination and a consequent reduction in its reflectance. It should be stressed that the presence of smoke and vapours makes many of the techniques to be described in this chapter difficult to apply, and that it is sensible to make provision for the regular cleaning of mirror surfaces where the method depends on their reflectance being high. Nevertheless, even if only two reflections contribute to the signal, and ρ_M is as low as 0.9, the effective emissivity of a surface with ϵ_λ equal to 0.5 is increased to 0.83, a substantial and worthwhile improvement. Moreover, the sensitivity of the radiation thermometer reading to changes in ϵ_λ is significantly reduced.

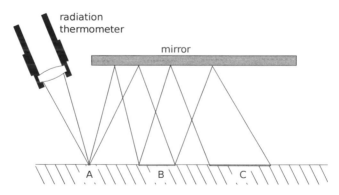

Figure 8.3: The effect of placing a mirror above a hot specularly reflecting surface.

The analysis so far has taken the surface to be a specular reflector. When the surface is a perfectly diffuse reflector a similar effect occurs, but the increase in the effective emissivity is usually much smaller because of the geometrical losses after scattering at the surface. The situation is more complex, and no attempt will be made here to obtain a mathematical solution. The first additional component (Figure 8.4) arises from radiation emitted within the area B and reflected from some part of the mirror onto the radiation thermometer target area A. There it is diffusely scattered, with a fraction $\omega_p \cos \theta / \pi$ entering the thermometer. The second component consists of radiation from any area of the surface which can be reflected onto B, after which it is diffusely scattered and a part may follow the same path as the first component. Unlike the specular case, the losses at each reflection from the surface contain a geometrical term containing the solid angle ω_M subtended by the mirror from the point of scattering within B.

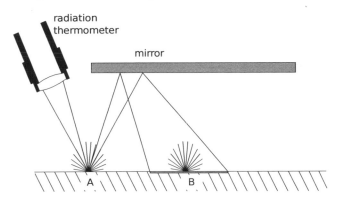

Figure 8.4: Increase of apparent emissivity with a mirror placed above a diffusely emitting surface.

In both specular and diffuse cases, the assessment of the improvement in apparent emissivity to be obtained with this simple technique is difficult. The first problem is that the complexity of the geometry increases rapidly as the number of reflections from the mirror increases, so that only an approximate value for the number and magnitude of the additional terms may be made. Secondly, the sources of the additional radiation spread out across the surface and it becomes difficult in practice to ensure that the assumptions on the uniformity of the surface conditions are maintained. One solution is to tilt the mirror towards the target area A. This brings the contributing areas of the surface towards and around A, but may limit the number of reflections which can be included. If the lower end of the mirror approaches the surface closely, the enclosed space forms a simple black-body cavity [22][23]. Clearly, this can only be achieved in practice when the surface temperature is relatively low and the system is free from dirt and condensing vapours.

When the tilt from the horizontal is equal to θ, only one additional source centred approximately at A is possible, although this again covers a greater area than A as the radiation thermometer is effectively defocused. The use of a spherical or concave mirror in this position (Figure 8.5) provides a symmetrical configuration with the reflected radiation originating at or close to the target area A. Under these conditions the effective emissivity of the surface may not be large, but it may still be possible to obtain useful information. If the surface is specular and we measure the ratio of the signal with the mirror reflection, $I_0 + I_1$, to that without it, I_0, we obtain from (8.2)

$$\frac{I_0 + I_1}{I_0} = 1 + \rho_M(1 - \epsilon_\lambda) \qquad (8.4)$$

and this enables $\epsilon\lambda$ to be calculated if the mirror reflectance is known. One of the advantages of this configuration is that the cold mirror shields the target area from thermal radiation from other sources which might otherwise enter the radiation thermometer. Radiation from other directions cannot be specularly reflected into the instrument. To measure I_0 alone, it is better to replace or cover the mirror with a non-reflecting surface rather than to remove it, so that stray radiation cannot affect the measurement. It may be noted at this point that, while the original use of mirrors was simply to increase the effective emissivity, the modern trend is to measure both the fundamental and the augmented signals, to give estimates of both the surface emissivity and the temperature.

The method is particularly appropriate for the characterisation of shiny metallic surfaces, as the largest signal ratios are obtained for materials with low emissivities. It would appear that the uncertainty in ϵ_λ should be low, as the method is simple and (8.4) is well behaved. Both the signal ratio and the mirror reflectance ρ_M may be determined with reasonable accuracy, so that the best results should be obtained when the emissivity is low and the required temperature correction large. In fact, the reliability of the results obtained depends critically upon the assumption that the surface behaves as a specular reflector, and in general some adjustment must be made for the presence of diffuse reflections.

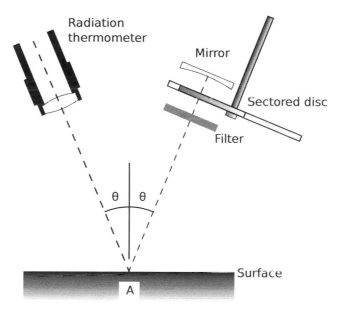

Figure 8.5: Measurement of emissivity and temperature for a specular surface with a single reflection from a spherical mirror.

To allow for a small amount of diffuse reflection, it is assumed that the reflectance of the surface may be split into components representing pure specular and diffuse (Lambertian) behaviour,

$$1 - \epsilon_\lambda = \rho_S + \rho_D \qquad (8.5)$$

that is, for an incident beam a fraction ϵ_λ is absorbed by the surface, ρ_S is specularly reflected at the same angle and the remainder ρ_D is diffusely scattered into a hemisphere. This assumption is not too unreasonable under near-specular conditions, but becomes increasingly inaccurate as the surface roughness and hence the diffuse reflectance increases. In practice, the relative proportions will depend on θ, particularly at large angles from the normal to the surface. The fraction f_D of diffuse reflection is defined by

$$f_D = \frac{\rho_D}{\rho_S + \rho_D} = \frac{\rho_D}{1 - \epsilon_\lambda} \qquad (8.6)$$

We also assume in this case that there are no major external sources which could be scattered into the radiation thermometer from the diffuse component of the reflectance.

The signal I_0 from the radiation emitted directly into the thermometer is again given by (8.1). The additional component I_1 produced by the radiation from A collected by the mirror, returned to A in the solid angle ω_m and reflected or scattered into the acceptance cone solid angle of the instrument ω_p is

$$I_1 = k\epsilon_\lambda L_\lambda(\lambda, T) A \cos\theta \omega_m \rho_m \left(\rho_S \frac{\omega_p}{\omega_m} + \rho_D \frac{\omega_p \cos\theta}{\pi} \right)$$

and hence their ratio is

$$\frac{I_1}{I_0} = \rho_m \left(\rho_S + \rho_D \frac{\omega_m \cos\theta}{\pi} \right)$$

$$= \rho_m(1 - \epsilon_\lambda) \left[1 - f_D \left(1 - \frac{\omega_m \cos\theta}{\pi} \right) \right] \qquad (8.7)$$

It might appear that it would be possible to estimate f_D by repeating the measurement of the signal ratio with a different value for ω_m, either by placing different apertures in front of the mirror or by moving it towards or away from the target area A. However, the term involving ω_m is relatively small, and the change in the signal ratio would not exceed one or two per cent, which is inconveniently close to the measurement accuracy.

8.4.1 Extension to hemispherical reflectors

The technique described above is intended to be applied to specularly reflecting surfaces. To obtain a worthwhile increase in the apparent emissivity with diffusely reflecting materials, the solid angle of the mirror viewed from the target area must be large. Ideally, we would wish to achieve hemispherical illumination of the surface with the reflected radiation. The simplest and most logical way to do this is to employ a hemispherical mirror. A radiation thermometer based upon this principle has been in use for many years [24]. The principle is shown in Figure 8.6. It consists of a hemispherical reflector coated with polished gold or alternatively platinum to give high reflectance over a broad wavelength range. In use, the device is placed close to the target area; thermal radiation from the surface is largely returned to it and establishes a radiation field within the intervening space which approximates that inside a black-body at the true temperature of the surface, even though the mirror remains cold.

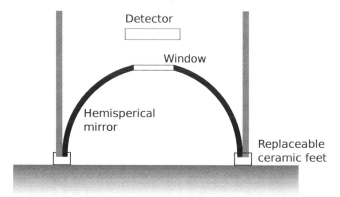

Figure 8.6: The sensing head of a hemispherical or gold cup radiation thermometer, showing the detector aperture and window, the replaceable ceramic feet, the photodetector and the hemispherical gold-plated mirror.

The radiation density is sampled through a small aperture behind which is placed a photodetector appropriate to the temperature range of interest. The aperture is sometimes closed by a window, to protect the photodetector and to enable contamination to be removed simply by cleaning the mirror and window, which are readily accessible. The apertures in front of the detector limit the target area to a circle of radius about one-third of that of the hemisphere. Typically, the hemisphere is of 25 mm radius, while that of the window aperture would be about 5 mm. For operation in very confined spaces, for example, for monitoring temperatures during welding inside or between boiler tubes, a very small mirror may be combined with fibre optics which transmit the thermal radiation to a remote detector. This system may be placed within a tube which protects the optical fibre and which carries water cooling and an air purge to limit contamination to the mirror. By keeping the detector away from the head, interference from the electrical noise accompanying the welding process may be considerably reduced.

To achieve a reasonable approximation to black-body conditions, the separation between the rim of the mirror and the surface must be small. As some radiation will inevitably be lost through the gap, small ceramic spacers may be placed at a few points around the rim to improve the positional reproducibility. It should be stressed that the reflecting mirror techniques under discussion here lose some of the advantages of more conventional radiation thermometric methods. Not only may the device itself be heated or contaminated by the hot surface, but its presence may affect the actual temperature of the surface, especially if the material is of low thermal conductivity. Although the radiative and conductive heat losses are reduced by the presence of the mirror, so that the surface temperature rises towards that in the interior of the material, the equilibrium time may be long, and the reading from the radiation thermometer will correspond to some intermediate and indeterminate state. To avoid these problems, the thermometer is usually placed close to the surface for a short time, sufficient only to take the necessary reading.

The simple geometry of the hemispherical mirror enables a reasonable estimate of the improvement in the effective emissivity of the surface to be obtained without difficulty. We assume that the separation h between the rim of the mirror and the (flat) surface is small, so that the target area viewed by the detector is at the centre of the hemisphere of radius R. Initially we shall consider that the detector is focused on a small target area at this point. Rays emitted from the centre, apart of course from those lost through the gap at the rim or through the detector aperture of radius r, will be reflected back to their point of origin.

If the initial angular distribution of radiation emitted by the surface is Lambertian, the proportion F of radiation striking the mirror, that is, not lost through the gap at the rim and through the detector aperture, is

$$F = 1 - \frac{r^2 + h^2}{2R^2} \tag{8.8}$$

Some of the radiation passing through the detector aperture gives rise to a direct signal I_0:

$$I_0 = k\epsilon_\lambda L_\lambda(\lambda, T) A \omega_p \tag{8.9}$$

The proportion returned to the surface is $\rho_M F$ where ρ_M is the reflectance of the mirror. Of the radiation reaching the surface, we assume that a fraction ρ_D is then diffusely reflected with an ideal Lambertian distribution; some of this will produce an additional contribution to the total signal I_T. While a fraction F of this diffuse component will be lost at the second reflection from the mirror, it is assumed on the other hand that the specular component ρ_S returns without loss and does not contribute to the detected signal. The remainder is absorbed by the surface according to (8.5). The component ρ_S should not be interpreted as an ideally specular reflection; the model attempts to allow for the complex angular distribution of the reflected rays by providing this parameter as a measure of those which do not contribute to the total signal, while ρ_D represents the fraction that does.

As the inter-reflections between the mirror and the surface continue, the angular distribution of the radiation taking part remains to a large part unchanged, and we may write for the total signal

$$I_T = I_0 \left\{ 1 + \rho_D F \rho_M \sum_{N=0}^{N=\infty} [\rho_M(\rho_D F + \rho_S)]^N \right\}$$

$$= I_0 \frac{1 - \rho_M \rho_S}{1 - \rho M(\rho_D F + \rho_S)} \tag{8.10}$$

Substituting for the surface reflectances in terms of f_D and the emissivity from (8.5) and (8.6) gives for the effective emissivity ϵ' of the surface

$$\epsilon' = \epsilon_\lambda \frac{1 - \rho_M(1 - \epsilon_\lambda)(1 - f_D)}{1 - \rho_M(1 - \epsilon_\lambda)(1 - f_D + f_D F)} \tag{8.11}$$

The signal is determined principally by the normal rather than the hemispherical emissivity. Any major differences between the two, and the deviations from a Lambertian distribution, occur at large angles from the normal, where the contribution to the total signal is small.

The improvement in the apparent emissivity of the surface is shown in Figure 8.7, for a typical hemispherical radiation thermometer with a clean gold mirror surface. It will be seen that there is a substantial approach towards black-body behaviour even for surfaces with relatively low emissivity, but that the value attained depends quite critically on the assumption of diffuse reflection ($f_D = 1$). Indeed, for completely specularly reflecting surfaces ($\rho_D = 0$) no improvement at all is obtained, as there is then no mechanism for rays being reflected between the mirror and the surface to be scattered towards the photodetector. A small improvement in this respect may be achieved by placing the detector aperture asymmetrically [23]. Otherwise, only diffusely reflected radiation can contribute to the additional signal.

In general of course, the detection system is not focused at the surface, but receives radiation from a much greater area. This has little effect upon the diffuse component of the signal. For specular surfaces a larger target area results in a significant increase in the effective emissivity.

The calculation of the effective emissivity for a broad target area in the specular case is not analytic. For diffuse surfaces, the directional information in the incident beam is completely lost on collision with the surface and the angular distribution restored to its initial form, usually taken to be Lambertian. The same set of angle factors which describe the distribution of the radiation therefore apply after each reflection at the surface. For specular surfaces, each ray must be followed through many reflections whose number may depend critically upon its initial direction. Fortunately there is a simple ray-tracing procedure which may be used to simplify the calculation considerably.

We consider a ray passing through the centre of the detector aperture at an angle θ. If we back-track along this ray (in the opposite direction to the arrows) through its reflections within the hemisphere, its various components originate at the points where it strikes the emitting surface. To simplify the geometry of the

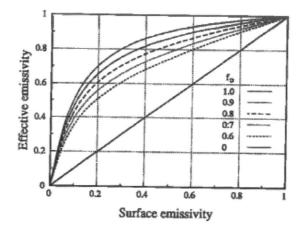

Figure 8.7: The effective emissivity of a surface covered by a hemispherical mirror with the detector focused at the centre calculated for $R = 25\,mm$, $r = 5\,mm$, $h = 3\,mm$ and $\rho_M = 0.92$ for different values of the parameter f_D.

system, we draw the image of the hemispherical mirror in the surface (Figure 8.8), and allow the ray to pass undeflected through the surface plane. In reality, all of the rays in the lower hemisphere should be folded up into the volume of the upper hemisphere. The problem is essentially two-dimensional since reflections at the mirror and surface remain in the same plane as the exiting ray.

Simple geometry shows that the angle α of the radii to the points of contact of the ray and the mirror (and its image) starts at 90° and increases by $(180 - 2\theta)$ for each reflection. The coordinates of the points of contact are $(R\cos\alpha, R\sin\alpha)$ when the origin O is at the centre of the hemisphere. To take account of the reflective losses, the backtracking ray is given an initial intensity of unity which is reduced by ρ_M and $(1 - \epsilon_\lambda)$ for each reflection at the mirror and the surface, respectively. The intensity of the exiting ray is then the sum of the components produced for each passage of the backtracking ray through the surface which has magnitudes equal to the product of the intensity of the incident ray and the emissivity of the surface at the angle of incidence. The latter is equal to $(2n + 1)\theta$, where n is the number of previous reflections at the mirror. In the calculations presented later, the dependence of the emissivity upon angle has been neglected. It should be noted that the ray sometimes does not pass through the surface after reflection at the mirror, but strikes the mirror again on the same side of the surface plane. This occurrence is detected by noting that the sign of the vertical coordinate $R\sin\theta$ does not change between reflections. Clearly there is no contribution to the intensity of the exiting ray in this case.

The process is terminated when the reflection coordinates indicate that the ray originated from within the detector aperture (or its image) or has entered the hemisphere through the gap at the rim. It is assumed that no radiation can enter or be reflected back through the aperture; rays through the gap are allowed, but

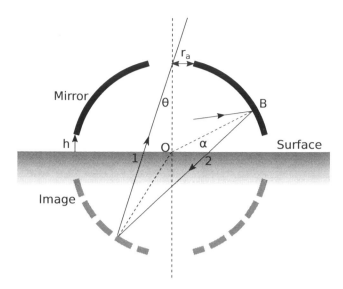

Figure 8.8: Ray-tracing method for determining the effective emissivity of a hemispherical mirror over a specularly reflecting surface.

no further reflections are considered to take place outside the hemisphere. The calculation is also stopped if the local intensity of the ray falls below a threshold value, for example 0.001. The backtracking procedure is repeated for a suitable range of values for θ to give $\epsilon'(\theta)$. This function is not continuous, and a reasonable density of values is required to obtain acceptable accuracy. To calculate the overall effective emissivity ϵ' the function must be integrated or summed over the range of angles accepted by the detection system, up to a maximum value θ_M:

$$\epsilon' = \frac{\int_0^{\theta_M} \epsilon' \sin\theta \cos\theta d\theta}{\int_0^{\theta_M} \sin\theta \cos\theta d\theta} \tag{8.12}$$

A plot of this against surface emissivity is shown as curve B in Figure 8.9 for typical parameters for a gold cup radiation thermometer. It will be seen that the effective emissivity is in this case not too different from that for a diffuse surface.

We may conclude that the gold cup radiation thermometer should work satisfactorily with flat, clean and accessible surfaces so that the instrument head may be placed close to them briefly to take a reading. The temperature obtained corresponds to a high effective emissivity, and is relatively insensitive to changes in the real emissivity of the surface or its roughness. The thermometer may be calibrated for use with a given material, giving some correction for the residual difference from black-body conditions. Calibration also allows for deviations of the mirror reflectance from the expected value due to defects in its preparation or contamination in use.

Masking the mirror with a black screen allows the ratio of I_0 from (8.9) to the total signal to be determined, which would enable an estimate of ϵ_λ to be made. There are two problems with this proposal: first, the practical one of effectively

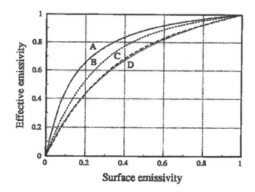

Figure 8.9: Calculated effective emissivities A, B – for an unfocused hemispherical radiation thermometer ($\theta_M = 18°$); C, D – for a modified spherical mirror radiation thermometer, for diffuse and specular surfaces, respectively.

masking the mirror without changing the separation h or obscuring the field of view of the detector and, secondly, of allowing for the fact that the signal ratio depends upon the surface characteristics through the factor f_D. The analysis carried out above suggests a means for minimising both. The effective emissivity for diffuse surfaces may be controlled by changing the separation, while that for specular surfaces by altering the detector aperture radius r. It appears possible to design a spherical mirror system which responds equally to diffuse and specular surfaces, as shown in Figure 8.10, and has an increased working distance from the surface to enable screening of the mirror to be more easily effected. In Figure 8.9 such a device can be seen although having a smaller enhancement of emissivity would be less dependent on surface condition.

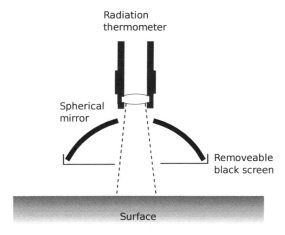

Figure 8.10: Modified spherical mirror radiation thermometer with a reduced response to the degree of surface roughness.

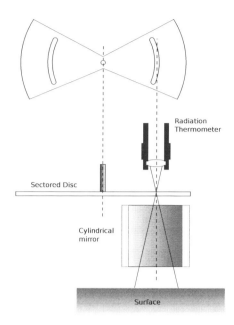

Figure 8.11: Sectored disc and cylindrical mirror system.

8.4.2 Use of cylindrical mirrors

The placing of a screen between the surface and a hemispherical or a spherical mirror in order to measure the fundamental signal I_0 gives rise to a number of practical problems. The use of a cylindrical mirror [25] enables these to be largely overcome, and both I_0 and I_T measured quickly without altering the state of the surface. The device is shown in Figure 8.11. A motor-driven sectored disc rotates above a cylindrical mirror, alternately opening and closing the upper aperture. The inner surface of the mirror and the lower surface of the disc are gold-plated and polished to give a highly reflecting surface. When the cylinder is open, the detection system accepts only radiation directly from the target area; reflections from the walls of the cylindrical mirror fall outside the field of view. When it is closed by the disc, these rays are reflected back onto the surface, and may be diffusely scattered or reflected through a slot cut in the sectored disc into the detection system. The slot is sufficiently large that it does not obscure the field of view of the detector. As the cylindrical mirror is stationary, and the sectored disc rotates in a time much less than the thermal time constant of the surface, the measurements of I_0 and I_T are obtained under closely similar conditions.

Practical results have been given by Iuchi [25], referred to as the modulation-aided (MA) method. The mirror was formed from a water-cooled brass cylinder of 50 mm diameter and 150 mm length. The width of the aperture slot in the sectored disc was 5 mm, and a lead sulphide detector was used. By making a number of simplifying assumptions, it was shown that the ratio of I_T and I_0 could be written in the form

$$\frac{I_T}{I_0} = \frac{\alpha + 1}{\alpha + \epsilon} \tag{8.13}$$

where α is a parameter which should depend upon the geometry of the system and the proportions of diffuse and specular reflectance of the surface, but not upon its emissivity. To check this assumption, samples of different metals and alloys were heated in a furnace and their temperatures measured with a type K thermocouple. The values of I_0 and I_T obtained from the thermometer enabled the emissivity and the factor α to be determined. The published results are shown in Table 8.1. It will be seen that α is essentially constant for some materials over a wide range of emissivity values, implying that the roughness did not change as the surface oxidised and the emissivity increased. It was also shown that α was closely related to the surface roughness parameter σ in the range 0.1 µm to 10 µm by an equation of the form

$$\alpha = 0.10 \ln \sigma + 0.52 \tag{8.14}$$

An analysis of the performance of this type of thermometer appears more complicated than in the case of a hemispherical mirror, as the rays from the surface may be reflected several times from the cylinder before returning to the emitting surface. As a result, rays from a given point are redistributed over the area covered by the mirror (and even just outside), depending upon the angle of emission. Nevertheless, the simple analysis presented here enables a semi-quantitative understanding of the behaviour of the device to be established. As a first step we calculate the reflective loss factor ρ' for a Lambertian source on the surface. Unlike the situation for the hemispherical mirror, this factor includes both the losses from the reflectance ρ_M of the mirror (assumed to be the same for the cylinder and the sectored disc) and the geometrical losses through the gap h at the rim and through the detector aperture of radius r.

Table 8.1: Characteristics of metals and alloys

Material	Temperature	Emissivity	α
Dull cold rolled steel	200 °C to 600 °C	0.19-0.81	0.53 ± 0.04
Stainless steel, type A	300 °C to 900 °C	0.20-0.74	0.25 ± 0.02
Stainless steel, type B	300 °C to 900 °C	0.21-0.73	0.34 ± 0.02
Polished aluminium	300 °C to 500 °C	0.06-0.08	0.38 ± 0.02
Shot blasted aluminium	300 °C to 500 °C	0.30-0.38	0.75 ± 0.04

A reasonable approximation to ρ' may be obtained for radiation from the central point O of the covered surface (see Figure 8.12). It may be shown that the values at other points up to the edge of the mirror do not differ by more than a few percent as long as the height H of the cylinder is greater than its radius R. Rays from the centre may be divided into zones characterised by the number of reflections from the mirror required to return a ray to the surface. Each ray is reflected once and only once from the top of the cylinder formed by the sectored disc, unless immediately lost through one of the gaps. As the angle of emission θ from the normal to the surface increases, the number of reflections increases. The limiting angle θ_N for each zone corresponds to a ray which just touches the bottom rim of the mirror, that is, for N reflections,

$$\tan\theta_N = \frac{(2N-1)R}{2H+h} \quad : \quad \theta_0 = 0 \tag{8.15}$$

Rays which are emitted at an angle greater than a maximum value θ_M do not strike the mirror at all, and are lost. The limiting angle which is smaller or equal to this defines an integer N_M related to the maximum number of reflections allowed, and given by

$$\tan\theta_M = \frac{R}{h} \quad : \quad N_M = Int\left(\frac{H}{h}\right) + 1 \tag{8.16}$$

Integrating for Lambertian radiation over the solid angles corresponding to the various ranges of θ gives

$$\rho' = \sum_{N=0}^{N_M+1} \rho_M^{N+1}(\cos^2\theta_N - \cos^2\theta_{N+1})$$

Rearranging and allowing for the fact that the maximum angle for rays to enter the mirror is θ_M,

$$\rho' = \rho_M\left[1 - (1-\rho_M)\sum_{N=1}^{N_M}\rho_M^{N-1}\cos^2\theta_N - \rho_M^{N_M}\cos^2\theta_M\right] \tag{8.17}$$

The cosine terms may be expanded as functions of the dimensions and position of the mirror using (8.15) and the identity

$$\cos^2\theta = \frac{1}{1+\tan^2\theta}$$

To this point, the effect of the detector aperture on the loss factor has been neglected. This may be calculated in a similar fashion, but for mirror geometries suitable for the present application amounts to a reduction in ρ' of less than 1 %. The detector aperture or slot is usually of the order of 5 mm across for a mirror radius R of 25 mm. It should be noted that for realistic values of ρ_M, R, H and h, the reflective loss factor ρ' is quite low, between 0.5 and 0.7, mainly as a result of the multiple reflections experienced by rays emitted at large angles.

The calculated values may be used to estimate the effective emissivity for perfectly diffuse surfaces, from

$$\epsilon' = \frac{\epsilon_\lambda}{1 - \rho'(1-\epsilon_\lambda)} \tag{8.18}$$

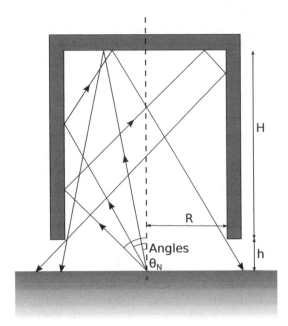

Figure 8.12: Calculation of the reflective loss factor for diffuse radiation in the closed cylindrical mirror.

but in the general case we would wish to include surfaces which show a mixture of diffuse and specular characteristics. We again introduce the assumption that we may represent this behaviour by a combination of perfectly diffuse and perfectly specular reflections. To calculate the effective emissivity for specular reflections the backwards ray-tracing technique is again employed, using the geometry shown in Figure 8.13. Rays passing through the centre of the detector aperture at an angle θ are traced backwards to their point of origin. Those striking the surface are continued through to the image of the mirror as shown. If the coordinate origin is taken at the centre of the covered surface O, the y-coordinates range from $(H+h)$ to $-(H+h)$. If the x-coordinate exceeds $\pm R$ then a reflection from the mirror wall has taken place. This could be handled by reversing θ at this point, but we may take advantage of the symmetry of the system by adding or subtracting $2R$ from the x-coordinate to translate the ray to the opposite surface of the mirror while keeping θ constant. This simplifies the programming required to obtain the solution.

When the ray has reached the lowest point of Figure 8.12, the image of the inner surface of the sectored disc, the procedure is repeated by re-introducing it at the top with the same x-coordinate and direction θ. As with the hemispherical mirror, the process is terminated when the ray passes through the gap between the mirror and the surface or through the detector aperture. The intensity of the backtracked ray is given an initial value of unity, and reduced by ρ_M and ρ_S for each reflection at the mirror and at the surface, respectively. The use of ρ_S allows for the scattering of a proportion of the incident ray, so that it no longer takes part in the specular reflections being tracked. When the intensity falls below an arbitrary threshold level, 0.001 for example, the process is also terminated.

Each intersection of the backtracked ray and the surface represents an additional contribution to the effective emissivity of magnitude equal to the product of ϵ_λ and the incident intensity. For cylindrical mirror systems, the angle θ is constant and not greater than a few degrees. The effective emissivity is therefore determined by the normal spectral emissivity of the surface and no allowance need be made for the angle of emission. The effective emissivity for the mirror as a whole is obtained by integrating over the permitted values of θ to cover the field of view of the detector. In the case of the apparatus shown in Figure 8.11, the detection system is focused at the aperture in the sectored disc, and the maximum angle is determined by the optical system employed. In the calculations which follow, the target area on the surface is taken as the central region of the covered surface, out to a radius of $R/3$. It should be noted that the effective emissivities obtained are quite dependent upon the field of view assumed.

To allow for surfaces which show a mixture of specular and diffuse characteristics, we add an idealised diffuse component to the specular reflection process described above. At each reflection from the surface, this is reduced in intensity by the fraction absorbed by the surface $(1 - \epsilon_\lambda)$ and at each reflection from the mirror by the reflective loss factor ρ'. Reflection at the surface adds a component to the effective emissivity equal to the product of the incident diffuse intensity and the diffuse reflectance ρ_D. The calculation is allowed to proceed until both the specular and the diffuse intensities have fallen below the specified threshold level. The results represent the effective emissivity of the surface as a function of the emissivity ϵ_λ and the diffuse fraction f_D of the reflectance (8.6). For comparison with the results of Iuchi and colleagues, they have been modified to give the parameter α which is plotted in Figure 8.14. It will be seen that the coupling between the emissivity and the surface roughness is not completely removed, but causes the curves for α to slope in a complex fashion. However, the calculated curves are not inconsistent with the experimental results given in Table 8.1. For shiny surfaces with a low value of f_D with α around 0.2 to 0.3, the dependence on emissivity is weak over a wide range of values. It is possible also that the effects of oxidation which lead to higher emissivities also increase the roughness of the surface and hence the value of f_D. This would have the effect of reducing any changes in α. In any case, α should be regarded as an empirical parameter which describes the behaviour of a given surface, and its numerical range must be established by experimental measurements rather than assumed or modelled by approximate calculations.

8.5   USE OF AUXILIARY SOURCES

A rather more versatile method of obtaining information on the surface characteristics is to direct an additional source of radiation onto the target area. Measurements of the reflected or scattered radiation enable estimates of the reflectance and hence emissivity to be made. The contribution from the source may be distinguished from

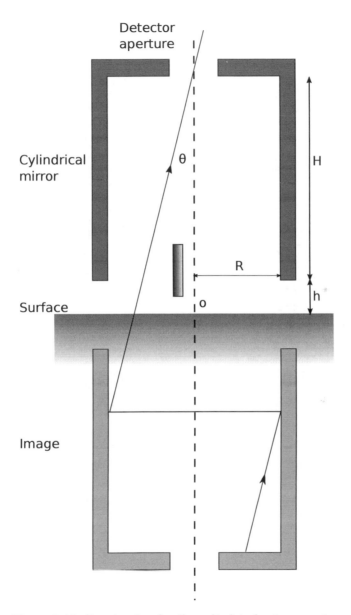

Figure 8.13: Ray-tracing for the cylindrical mirror system.

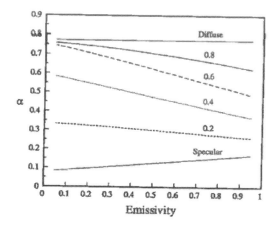

Figure 8.14: The parameter α calculated for values of f_D varying from specular ($f_D = 0$) to diffuse. ($f_D = 1$).

the thermal emission from the surface either by varying the source temperature, which is usually a slow process, or by chopping it and measuring the a.c. component of the thermometer signal. In the general case, the emissivity at an angle θ is calculated from the hemispherical reflectance

$$1 - \epsilon_\theta = \rho_H(\theta)$$

For some surfaces it is possible to employ (8.5), that is, to separate the diffuse and specular components of the reflectance. The effects of the limitations inherent in this assumption should be considered, especially for nearly diffuse surfaces. In addition, if the radiation from the source is polarised, or if the radiation thermometer is polarisation sensitive, the reflectances for the two polarisation states need to be known, particularly for the specular component.

Clearly, a complete evaluation of the surface characteristics would be a time-consuming and complex process, and most techniques to be described in this section attempt to introduce some simplifying assumptions. At near-normal incidence, for example, polarisation of the emitted and reflected radiation is small. At large angles of incidence, the radiation is polarised, but most smooth surfaces appear nearly specular. It is of course important to establish the conditions under which the assumptions made are valid, or the magnitude of the corrections required if they are not.

Three types of sources have been used for this work. Small, electrically heated black-bodies are ideal in that the spectral radiance is well known if the temperature is monitored, and that any radiation reflected into them is completely absorbed. Apertures may be required to reduce the effective diameter of the beam produced, should stray radiation from the source itself appear to be a problem. In practice they are rather inconvenient; together with the associated power supplies and temperature measurement and control system they form a cumbersome

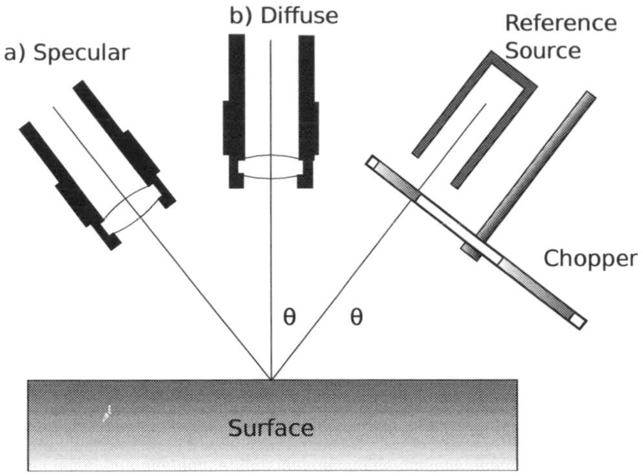

Figure 8.15: Determination of the surface characteristics and temperature using near-normal irradiation of the target area with chopped radiation from a reference source.

system with a tediously long response and settling time. As an alternative, heated plates with coated surfaces to give high emissivity are simpler to construct and faster in response. They may also provide some screening of the target area from stray radiation. These advantages are gained at the expense of precise knowledge of the spectral radiance generated, and a broad but not hemispherical illumination of the target area. Lasers, on the other hand, provide a well-collimated beam of monochromatic radiation at a precisely defined wavelength. The intensity is high, so that high signal-to-noise ratios may easily be achieved, and may usually be precisely stabilised with an electronic control system. The fact that the output is usually linearly polarised is sometimes a positive advantage. If not, the polarisation state may be converted to that required with a few simple optical components.

8.5.1   Methods with near-normal irradiation of the surface

The simplest technique for determining the surface characteristics is to illuminate the target area with a reasonably well-defined beam at a small angle from the normal to the surface, and to measure separately the specular and diffuse components of the reflected radiation. The arrangement is shown diagrammatically in Figure 8.15. At small angles, the differences between the polarisation states may be neglected. The beam from the black-body or laser source is chopped to distinguish it from the thermal radiation from the surface, which is measured along the direction of the specular reflection. When the radiation thermometer is moved away from this position, the a.c. signal corresponds to the diffuse component of the reflected radiation. In principle, it should be necessary to measure the distribution of the diffuse component at all angles, but if the surface is mainly specular it may be sufficient to measure its intensity normal to the surface, and to assume that Lambert's law is applicable.

The measurements yield three normalised signals. The d.c. component at the angle of specular reflection represents the thermal emission from the surface,

$$I_s = \epsilon_0 L_B(\lambda, T) \tag{8.19}$$

while the magnitude of the a.c. component is the specularly reflected signal from the reference source which has an effective radiance $L_B(\lambda, T_R)$, with a small contribution from diffuse reflection:

$$I_\theta = L_B(\lambda, T_R)(\rho_S + \rho_D \omega_P \cos \theta) \tag{8.20}$$

In the normal position, the a.c. signal is

$$I_n = \rho_D L_B(\lambda, T_R) \omega_P \tag{8.21}$$

If the radiance temperature T_R and the solid angle ω_P are measured directly, then these equations may be used to obtain ρ_S and ρ_D. The emissivity ϵ_θ is then calculated from (8.5), enabling the true temperature T of the surface to be estimated from (8.19). Alternatively, the surface may be replaced by a mirror with a known reflectance, in which case a measurement of I_θ enables $L_B(\lambda, T_R)$ to be determined. If the mirror is then replaced by a perfectly diffuse surface of known reflectance, I_n gives the acceptance angle of the radiation thermometer, ω_P. Although this procedure may appear to be time-consuming, it should not be necessary to repeat it frequently, as the source and thermometer characteristics should be stable. For surfaces which have a low fraction f_D of diffuse reflection, the diffuse components are small and ω_P is not required with any great accuracy.

The main problem with this technique is that the geometry of the system must be well controlled, so that the signal S_θ contains all of the specularly reflected radiation. This limits the permissible variations in the vertical position of the surface and in its tilt. These problems may be eased by allowing some divergence of the beam from the source, and illuminating a larger area than that required by the radiation thermometer. The need to move the thermometer between the two positions may be avoided by employing two detection systems. It is of course necessary to correct the readings obtained for differences in acceptance angle and operating wavelength so that the normalised signals may be inserted in the equations above. The critical assumption made is that the diffuse component of reflectance obeys Lambert's law; this is likely to be the case for microscopically rough surfaces where the specular component is small. For mainly specular surfaces, the diffuse component is in any case small, so that the resultant error is unimportant. Between these two extremes, however, failure of the assumption may give rise to significant errors, and an experimental check should be regarded as mandatory.

8.5.2 Multi-wavelength methods

To avoid the need to make absolute measurements of reflectance, a number of methods have been developed which attempt to make use of the much simpler measurement of the relative reflectances at two or more wavelengths. Gardner and Jones

[26] made the assumption that the variation of the total hemispherical reflectance with wavelength was the same as that of the specular reflectance ρ_S, which is of course much more easily measured. This is equivalent to stating that the diffuse and specular reflectances vary in the same way with wavelength. In this case, the emissivity may be written in the form

$$1 - \epsilon_\theta = G\rho_S$$
$$\text{where } G = \frac{1}{1 - f_D}$$

and G is treated as a constant. By measuring at several wavelengths the normalised signal

$$I_0(\lambda) = [1 - G\rho_S(\lambda)]L_B(\lambda, T)$$

and ρ_S from the amount of radiation specularly reflected by the target from a light source built into the instrument, G and T may be determined by non-linear least squares curve-fitting techniques.

Such an arrangement is shown schematically in Figure 8.16. Radiation from a quartz-iodine lamp is chopped by a rotating sectored disc and is incident on a hot surface at an angle. The specularly reflected component returns through a second outer ring of apertures on the disc so that it, and the thermal radiation from the filament, are chopped at twice the frequency. The beam passes through a rotatable filter wheel which selects wavelengths, and is then detected by a silicon photodiode. Phase-sensitive detection at the different frequencies enabled the two signals to be measured separately. Different wavelengths can be used and the spectral distribution of the quartz-iodine lamp measured by reflecting the beam with non-selective gold-coated mirrors directly into the detection system.

The assumption made is that f_D is independent of wavelength. However, it should be noted that quite small deviations may lead to large errors in the derived temperature. These arise as a result of the curve-fitting procedures used, in much the same way as the model errors for multi-wavelength thermometry discussed in Chapter 7.

One minor problem in the measurement of the specular reflectance is that of ensuring that the surface is at the correct angle to the beam of light. This becomes more acute as the surface becomes more diffuse in character. An alternative technique is to measure the ratios of the diffuse reflectances at three wavelengths. In this case there is no alignment difficulty. However, the analysis of the measurements relies on two assumptions: that the specular reflectance is independent of wavelength, and that the ratios of the diffuse reflectances are independent of angle. The second assumption enables the diffuse component to be expressed as the product of a wavelength-dependent factor and a constant geometrical term, and has been verified for aluminium, copper and stainless steel samples.

The technique developed by Kunz and DeWitt [27], on the other hand, is unusual in that no arbitrary assumptions are made about the optical properties of the

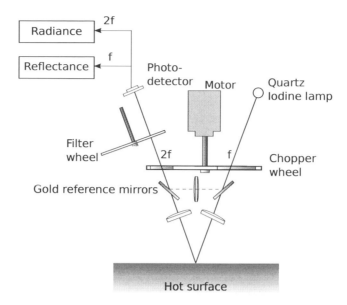

Figure 8.16: Emissivity correction by the measurement of the specular reflectance at different wavelengths where the modulation frequencies f and 2f refer to the reflected and emitted signals, respectively.

surface. It notes merely that the (small) temperature rise produced when part of the surface is irradiated, for example by a collimated laser beam, is proportional to the absorptance of the surface and to the power in the beam. If the surface is irradiated at different times by two wavelengths, λ_1 and λ_2, and the relative powers adjusted to give the same temperature rise, then the ratio of the absorptances, and hence from Kirchhoff's law that of the emissivities, is equal to the ratio of the beam powers, that is,

$$\frac{\epsilon_1}{\epsilon_2} = \frac{\alpha_1}{\alpha_2} = \frac{P_1}{P_2}$$

The emissivity ratio may be used directly to correct the reading of a ratio radiation thermometer working at the wavelengths λ_1 and λ_2 to give the true temperature of the surface.

Figure 8.17 illustrates the arrangement. To produce a temperature increase large enough to be accurately measurable, around $10\,°C$ to $20\,°C$, requires powers of $25\,mW$ to $50\,mW$. Moreover, it is preferable that a single laser should generate both wavelengths, so that the area and conditions of illumination of the surface are identical in both cases. The power output of the laser should also be stable or capable of being stabilised, as the technique requires a sequence of measurements which may take some time.

The temperature increase produced by the laser radiation is monitored with a broadband radiation thermometer designed to be sensitive to the relatively small changes (10 % to 20 %) in radiance produced. This instrument contains a filter such that it eliminates reflections and stray radiation from the laser at either λ_1 or λ_2.

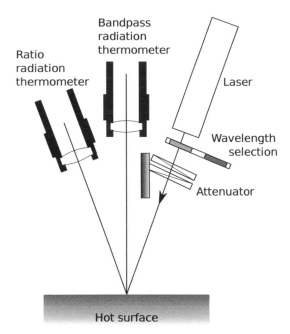

Figure 8.17: Measurement of the emissivity ratio at two wavelengths determined from the increase in temperature of the surface produced by the absorption of laser beams at those wavelengths.

The effectiveness of the filter can be checked by measuring the additional signal with the laser on and the tungsten strip cold, when the thermal signal is zero. The equilibrium temperature of the strip is measured with a ratio thermometer tuned to λ_1 and λ_2. The laser power at each wavelength is varied with an attenuator consisting of a pair of rotatable polarisers.

The technique has the distinct advantages that it can achieve high accuracy without making any critical assumptions about the characteristics of the surface. It is important of course that the conditions of illumination of the surface are the same at both wavelengths and there may be problems associated with surface temperature gradients, or the time required for the temperature increase to reach equilibrium, depending on the thermal diffusivity of the material.

8.5.3 Laser absorption radiation thermometry

An extension of the use of a varying heating source to modulate the surface is laser absorption radiation thermometry (LART) [28]. In this case a ratio method is used with two chopped or pulsed lasers at wavelengths λ_1 and λ_2 heating the surface and detectors operating at λ_2 and λ_1 measuring the thermally emitted source radiation.

Modulated heating of a surface from a laser operating at λ_t gives rise to an a.c. signal at a detector working at λ_l given by

$$I_{\lambda_l}^{\lambda_t} = R_t \epsilon_l(T) \epsilon_t(T) C_l C_t L_l(\lambda, T) P_l \Omega_t \Delta \lambda_t G_l(T) \tag{8.22}$$

where $\epsilon_l(T)$ is the emissivity at the laser wavelength, ϵ_t is the emissivity at the detecting wavelength, C_l and C_t are the transmission of the optical system at the two wavelengths, P_l is the laser power, R_t is the detector responsivity, Ω_t is the solid acceptance angle of the detection system, $\Delta \lambda_t$ is the detection bandwidth, L_t is equal to $\partial L_0 / \partial T$ with $L_0(\lambda, T)$ the black-body radiance and $G(T)$ is a function derived from the convolution of the source with the thermal response function of the target.

For LART, (8.22) applies to both heating at λ_1 and detection at λ_2 (l=1) and to heating at λ_2 and detection at λ_1 (l=2).

Where the whole of the modulated area lies within the field of view of the detection system the ratio G_1/G_2 is constant, and for an opaque surface

$$G_l(t) = \frac{1}{K}\left(\frac{D}{\omega}\right)^{1/2}$$

where K is the thermal conductivity, ω is the modulation frequency and D is the thermal diffusivity.

The emissivity terms cancel and so the ratio of two photocurrents $I_{\lambda 1}^{\lambda 2}$ and $I_{\lambda 2}^{\lambda 1}$ is taken, giving

$$Z = \frac{I_{\lambda 1}^{\lambda 2}}{I_{\lambda 2}^{\lambda 1}} = \frac{R_1 P_2 G_2(t) \Delta \lambda_1 L_0(\lambda_1, T)}{R_2 P_1 G_1(t) \Delta \lambda_1 L_0(\lambda_2, T)} \tag{8.23}$$

$$= k \frac{L_1(\lambda_1, T)}{L_1(\lambda_2, T)} \tag{8.24}$$

where k is a constant that can be determined by calibration at a known temperature.

The disadvantage of LART relates to the need to apply heating to the sample and as such it works better if the emissivity is not too low. For materials with emissivity at the 0.1 level an overwhelming large majority of the applied laser power is reflected rather than absorbed. This of course has consequences with the need to cope with reflected high power laser sources, possibly outside the visible spectrum. There is also a requirement that the emissivity is not temperature dependent over the temperature range that the target is modulated.

8.5.4 Illumination at large angles

Although illuminating a surface at an angle well away from the normal in general complicates the measurement problem, in that both the reflected and emitted radiation may well be distinctly polarised, there are some situations where an advantage can be gained. In particular, at high angles smooth surfaces become more nearly specular in behaviour, and this feature is accentuated at long wavelengths.

The latter point suggests that this approach may be suitable for the measurement of relatively low surface temperatures.

A simple configuration for such a measurement is shown in Figure 8.18. A black-body source illuminates the target area at a large angle, typically 75° to 80° to the normal. A narrow- or broadband radiation thermometer detects the radiation specularly reflected and also emitted from the target area. The field of view of the radiation thermometer should be smaller than the aperture of the black-body, so that it cannot receive radiation from any other source. Ignoring for the moment the effects of polarisation, the normalised signal may be written

$$I = \epsilon_\theta L_B(\lambda, T) + (1 - \epsilon_t heta) L_B(\lambda, T_R)$$

At low temperatures, the emissivity for a given geometrical arrangement will be stable and not change greatly with the surface temperature. It may be determined easily in preliminary experiments for a particular sample, for example by measuring the surface temperature T with a thermocouple, and then recording the thermometer readings as the reference black-body temperature T_R is varied.

At the large angles involved the directional emissivity may be greater than the normal emissivity, improving the signal-to-noise ratio considerably .The increase enables the surface temperature to be determined with significantly lower uncertainty.

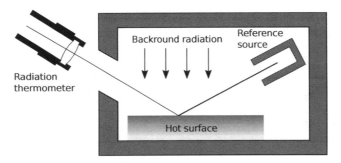

Figure 8.18: Determination of the temperature and emissivity of a surface from the specular reflection from a calibrated reference source.

The method assumes that, under the conditions given, the surface is completely specular. A diffuse component will of course produce errors, especially if stray or background radiation can then be scattered into the thermometer. If this is allowed for in the measurement equation, then

$$I = \epsilon_\theta L_B(\lambda, T) + (1 - \epsilon_\theta)(1 - f_D) L_B(\lambda, T_R) + (1 - \epsilon_\theta) f_D L_B(\lambda, T_A)$$

where T_A represents the effective black-body temperature of the background radiation. If the proportion f_D of diffuse reflection is known, then changing the temperature of the black-body source gives useful information on the emissivity ϵ_θ. If the parameter f_D does not change for a given material, that is, if the heat treatment it is undergoing does not alter the physical structure of the surface, the technique should be capable of coping with quite large variations in the emissivity.

The application of this technique has been described in some detail by Iuchi [29], under the acronym 'TERM' – temperature and emittance measurement by a radiation method. Of particular interest are the measurements, shown in Table 8.2, of f_D for three metallic surfaces at three wavelengths. These were made in a separate chamber in which the background temperature T_A was low and could be neglected. The sample and black-body temperatures could be varied independently, and were in the range 200 °C to 400 °C. The angle of illumination was 56°. It will be seen that, for a given material and wavelength, f_D was essentially constant over the surface temperature range, although the emissivity in some cases varied very widely. It should be noted that f_D varied with wavelength, so methods that assume it is constant will not work in this case.

Table 8.2: Measured values of the fraction f_D of diffuse reflectance

Material	Wavelength µm	Temperature range °C	Emissivity range	Diffuse reflectance f_D
Cold rolled steel sheet	2.2	326–410	0.31–0.82	0.08±0.03
	5.0	221–430	0.14–0.29	0.03±0.01
	8.0	296–431	0.14–0.38	0.081±0.01
Bright stainless steel sheet	2.2	306–410	0.26–0.36	0.02±0.02
	5.0	221–430	0.15–0.18	0.01±0.01
	8.0	296–431	0.16–0.18	0.00±0.01
Shot-blasted aluminium plate	2.2	306–410	0.38–0.43	0.91±0.01
	5.0	221–430	0.41–0.46	0.85±0.01
	8.0	326–431	0.45–0.48	0.74±0.01

For cold rolled steel at temperatures between 280 °C and 420 °C, the uncertainty in the measured temperature was 5 °C with a thermometer wavelength of 2.2 µm. While the temperature uncertainties were, as is usually the case, least at the shortest wavelengths, the uncertainties in the emissivity, expressed as a percentage, were smallest at the longest wavelengths.

8.5.5 Illumination with heated plates

A simple method which takes advantage of an external source to measure the emissivity and temperature of a surface may be developed by replacing the mirror in the configuration shown in Figure 8.3 by a heated plate. The ability to change the temperature of the plate gives an additional degree of freedom in determining the parameters required. The plate temperature may be monitored by directing the radiation thermometer at its upper surface. If a second radiation thermometer is employed for this purpose, it is necessary to correct the readings for differences in the sensitivity, the operating wavelength and the field of view between the two systems. In the analysis which follows, it is assumed that this has been carried out to produce the normalised signals given.

In general, the plate is designed to have the same high emissivity on the upper and lower surfaces, although we will not assume that this is the case in practice and is constructed of high thermal conductivity materials, so that both surfaces are at closely similar temperatures. If the surface is assumed to be perfectly specular, the normalised signal from the surface, made up of the thermal emission from the surface and the reflected radiation from the plate, may be written

$$I_1 = \epsilon_t heta L_B(\lambda, T) + (1 - \epsilon_\theta)\epsilon_L L_B(\lambda, T_L)$$

This expression assumes that there is no contribution from multiple reflections between the plate and the surface. The normalised signal I_2 from the upper surface is given by

$$I_2 = \epsilon_U L_B(\lambda, T_U)$$

In these equations, the subscripts L and U refer to the lower and upper surfaces of the plate, respectively. The measurement of the surface temperature is usually made by adjusting the plate temperature until I_1 and I_2 are the same. If the radiance temperatures of the upper and lower surfaces of the plate are equal, and this is a simple matter to check experimentally of course, this null condition reduces to

$$L_B(\lambda, T) = \epsilon_U L_B(\lambda, T_U)$$

that is, the temperature of the surface is equal to the radiance temperature of the plate.

In use, it is advisable to make a number of experimental checks. First, the equality of the radiance temperatures of the two sides of the heated plate may be tested with the radiation thermometer by inverting the plate. It should be noted that the lower surface may be affected by its thermal environment, that is, it may be heated by radiation from the sample. This effect may change the reading of the thermometer, but need not give rise to a significant error if the temperature gradient across the plate is small. Secondly, the technique may be validated for a particular material by comparing the temperatures obtained with those measured with a thermocouple inserted just below the surface of a sample. In addition, by varying the temperature of the plate away from the null point, it is possible to calculate the specular reflectance of the sample from the signals I_1 and I_2. The emissivity of the sample, on the other hand, may be calculated from the radiance temperature measured directly by the radiation thermometer, with the plate cold, and the true temperature obtained from the thermocouple or from the null measurement with the plate. A comparison of the specular reflectance and the emissivity will indicate whether the assumption that the surface is perfectly specular is valid or not.

In general, there will be a diffuse component of the reflectance, and to allow for this we should modify the equation for I_1 to

$$I_1 = \epsilon_\theta L_B(\lambda, T) + (1 - \epsilon_\theta)(1 - f_D)\epsilon_L L_B(\lambda, T_L)$$
$$+ (1 - \epsilon_\theta)f_D\epsilon_L L_B(\lambda, T_L)(\omega_H/2\pi)$$

where ω_H is a measure of the solid angle subtended at the target area by the plate. As the target area is not illuminated hemispherically, the signal observed is less than that expected if the assumption of perfectly specular reflectance were correct.

8.5.6   Polarisation techniques

We have noted that existing theories of surface characteristics are in general of limited value in predicting the emissivity of a real surface, or the form of its variation with either wavelength or angle. However, it has been suggested that the Fresnel equations may be employed to determined the emissivity of smooth polished metallic surfaces. At angle of incidence of 45°, the components of the specular reflectance obey the simple relationship

$$\rho_S^p = (\rho_S^n)^2 \tag{8.25}$$

where the superscripts p and n indicate rays with planes of polarisation respectively parallel and normal to the surface. If the reflection is completely specular, then the normalised intensities of the two components are

$$I^p = 0.5(1 - \rho_S^p)L_B(\lambda, T)$$
$$I^n = 0.5(1 - \rho_S^n)L_B(\lambda, T)$$

so that their ratio is given by

$$\frac{I^p}{I^n} = 1 + \rho_S^n \tag{8.26}$$

Measurement of the ratio therefore enables the reflectances to be obtained, and from these the emissivity and the temperature of the surface to be calculated.

Although the polarisation of radiation emitted or reflected at large angles to the normal is usually a practical inconvenience, some techniques have been developed which make specific use of it to correct for the unknown emissivity of the sample. The technique of 'polarisation radiation thermometry' basic configuration is shown in Figure 8.19. The radiation emitted by and reflected from the target area passes through a rotating analyser placed in front of the detection system. The measured signal varies sinusoidally, passing through maxima and minima corresponding to polarisations parallel and normal to the surface plane. If the superscripts p and n indicate the polarisation state, then the maxima and minima correspond to the normalised signals

$$I^p = 0.5[\epsilon_\theta^p]L_B(\lambda, T) + \rho_S^p L_B(\lambda, T_R)]$$
$$I^n = 0.5[\epsilon_\theta^n]L_B(\lambda, T) + \rho_S^n L_B(\lambda, T_R)]$$

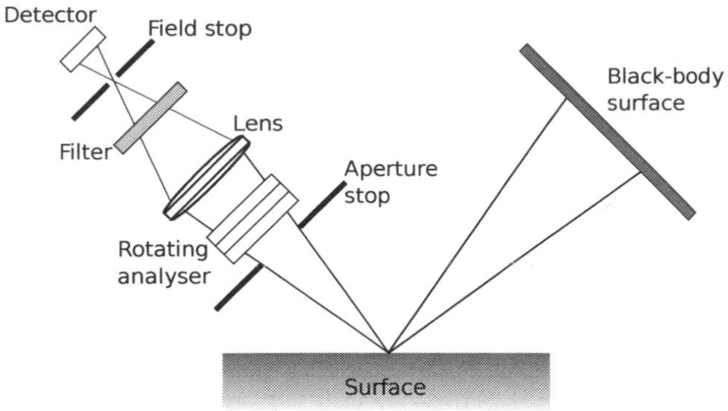

Figure 8.19: Polarisation radiation thermometer or 'polaradiometer'.

If the diffuse reflectance is independent of polarisation, then

$$\epsilon_\theta^p + \rho_S^p = \epsilon_\theta^n + \rho_S^n = 1 - \rho_D$$

from Kirchhoff's law, and the two signals are then equal when $T = T_B$ – whatever the degree of diffuse reflectance and the extent to which it is detected by the thermometer. The black-body temperature T_B is therefore varied until the a.c. component of the output signal is reduced to zero, and is usually monitored with a resistance thermometer or a thermocouple.

It is not of course essential to use a black-body source; a lamp, a heated strip or a laser may be used, but must conform to the requirements of the method. The beam must fully illuminate the target area and fill the field of view of the instrument. If the source is polarised to some extent, it should be depolarised. Some lasers are almost completely plane-polarised, and in this case the plane can be rotated, for example with a quarter-wave plate, avoiding the need for a polarisation analyser before the thermometer. Whatever the source, its effective temperature must be determined one way or another in order to obtain the surface temperature T.

The technique has a number of potential problems, apart from the need for the geometry of the system to be well defined. Although the null measurement appears convenient, the thermal time constant of the black-body source may make the approach to the balance point rather slow. The sensitivity of the method is proportional to the difference between ϵ^p and ϵ^n; if this is small, random or systematic errors near the balance point may convert into large temperature uncertainties.

8.6   MINIMISATION OF ERRORS FROM REFLECTED LIGHT

The effects of stray radiation from the surroundings or from extraneous sources have been discussed at a number of places in this book and especially in this chapter. In this section we shall add some remarks on the general problems of stray or background radiation and the methods which may be adopted to overcome them. The most important factors are of course the relative temperature of the unwanted source (or sources) compared to that of the target area and the emissivity of the surface to be measured. In addition we must consider their extent, that is, their apparent solid angle when viewed from the surface. Clearly, the target area may be screened with black absorbing baffles from the radiation from small sources, particularly if they are distant from the measurement region. When the source is large, for example the hot wall of a furnace containing the object whose temperature is to be measured, screening is more of a problem.

If the surface is specular in nature, then the major reflection into the radiation thermometer occurs at a well-defined angle, and a relatively small baffle or screen – preferably cooled – can be employed to remove it. If the surface is flat and the thermometer is sighted along the normal to the surface, the instrument itself can act in this way. If it is not possible to place screens in the required position, or if they heat up and become additional sources of radiation themselves, the temperature of the area viewed by reflection may be determined directly with a second radiation thermometer or with a contact thermometer, usually a thermocouple. The calculation of the correction is similar to that for illumination with heated plates, but requires that the diffuse and specular reflectances of the surface should have been previously measured.

For extended sources of stray radiation, and when the surface reflects diffusely, it is in general necessary both to reduce the additional component of the measured signal with appropriate screens and to make some correction for its presence. The screen should reduce the solid angle of the source 'seen' by the target area to a minimum. In practice the separation between the screen and the surface is often limited by the danger that a moving object in the furnace – a billet of steel, for example – may strike the screen with unfortunate results. An expendable or flexible skirt may be added to the lower rim of the shield to minimise the possible inconvenience. The underside of the shield facing the target area is coated black or oxidized to ensure that most of the stray radiation from outside the shield is absorbed. In addition, if the screen is maintained at a low temperature with respect to the surface to ensure that it does not itself act as a source of stray radiation, the radiative heat loss from the surface may lead to significant temperature gradients around the target area.

In furnaces at high temperatures, it may be impossible to maintain the screen at a sufficiently low temperature that it does not contribute to the thermometer signal. A correction may be derived by measuring the shield temperature with a thermocouple or a second radiation thermometer. An arrangement for doing this is shown diagrammatically in Figure 8.20. The temperature of the shield is determined with a radiation thermometer which is focused onto a small black-body cavity added to the upper surface of the shield. This avoids errors from temperature gradients in the vicinity. If the background radiation may be represented by a black-body source at a temperature T_F the signal from the radiation thermometer directed normally to the surface is

$$I_S = \epsilon_\lambda L_B(\lambda, T) + \rho_D[\epsilon_S \omega_S L_B(\lambda, T_S) + \omega_F L_B(\lambda, T_F)] \qquad (8.27)$$

where ρ_D is the diffuse reflectance of the target area. This equation is of course

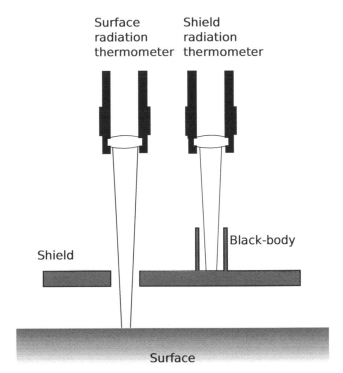

Figure 8.20: System for reducing the errors from stray radiation using a hot shield whose temperature is measured with a second radiation thermometer.

an approximation, and the surface constants should be treated as empirical parameters to be determined for each material and plant configuration. If the surface were perfectly diffuse and the surface, the shield and the furnace were at the same temperature, the surface radiation thermometer would see a source which would be close to a black-body at that temperature, so that

$$\epsilon_\lambda + \rho_D[\epsilon_S \omega_S + \omega_F] \approx 1 \qquad (8.28)$$

For a specular surface, the errors in this equation are too large for it to be useful in practice.

The emissivity ϵ_S of the screen and the solid angle ω_S it subtends at the target area are made as large as possible; conversely, the apparent solid angle ω_F for the stray radiation should be minimised. In practice, it is usually found to be considerably larger than that predicted purely from the geometry of the system, as a result of multiple reflections between the shield and the surface at near-grazing angles. Even so, the third term in (8.27) should be small. The magnitude of these terms are best obtained from preliminary tests in which the relative values of the operating temperatures can be controlled. The residual background signal may be estimated by adding cooled screening around the edge of the shield to remove it. The contribution from the shield may be assessed from the change in the signal I_S as the shield temperature T_S is varied, if necessary, by the addition of water cooling or extra heating. Finally ϵ_λ may be obtained by measuring the true temperature of the surface T by forming a small black-body cavity in it or by attaching a calibrated thermocouple. Under normal operating conditions, T_S is measured by the shield radiation thermometer (see Figure 8.20), while only a very approximate value for T_F is required. This may be estimated from the operating conditions for the furnace.

It should be stressed that this technique does not correct for changes in the emissivity or diffuse reflectance of the surface. These properties are assumed to be stable. If this is not the case, one of the emissivity correction methods described in the other sections of this chapter should be employed.

The meaning of 'measurement'

A.1 INTRODUCTION

Our definition of radiation thermometry employed the fundamental terms 'measurement', 'quantity' and 'temperature'. We shall pursue the question of their meanings in some detail, partly because it will eventually explain the relationship between temperatures measured with a pyrometer and those obtained with a thermocouple, for example.

Objects and events in the world around us show a variety of characteristics or properties. Some of these possess magnitude, that is, some of the objects have the property to a greater or lesser extent than others. These properties, for example size and weight, are called quantities; shape and colour, on the other hand, are generally considered to be qualities. One of the basic assumptions of science, and one reason why it has proved so successful in describing the physical world, is that it is possible to measure quantities.

Measurement is the assignment of a number according to established objective rules or procedures, in such a way that the number represents the magnitude of the quantity concerned. It should be distinguished from procedures which may superficially appear similar, but which involve the qualities of objects and some form of classification. For example, the Dewey decimal system assigns a number to a book so that it may be usefully stored and located in a library. Both the classification and the assignment are, however, to some extent arbitrary and subjective, and books with high numbers are in no way greater in any other respect than those with low numbers.

The act of counting is a very special form of measurement, to the extent that we would not in the scientific world employ the word 'measure' to describe it. In the first place it is exact; we can, barring blunders or incompetence, obtain an integer number which corresponds exactly with the number of trees in a wood, people in a room, and so on. Quantities generally have a continuous range of values and cannot be determined exactly. There is always some uncertainty associated with the measurement of a physical quantity, and this greatly complicates both the theory of measurement and its application in the real world. In addition, the units in which physical quantities are measured (about which more will be said later) characterise the quantity itself and are independent of the application. This is not to say that different units may not have been adopted for a given quantity, but the relationship between them is usually well known or even laid down by definition. The length of a molecule may be given in Angstrom units, that of a cricket pitch in yards, and the distance to the moon in kilometres, but all may be expressed if need be in metres. In counting, however, the 'units' are simply that which is being counted at the time – a tree, a person, etc.

A.2   CRITERIA FOR THE EXISTENCE OF A QUANTITY

In the physical sciences the fundamental concepts are well defined, so that measurements of the corresponding quantities are limited only by the uncertainties associated with the measuring procedures, which include the interaction between the measuring instrument and the quantity, and ultimately by the Heisenberg uncertainty principle itself. In other areas, the concept itself may be vague, and the results obtained will reflect this. This is often the case in the fields of economics, psychology and social sciences, for example. It is possible, therefore, to state the conditions necessary for the existence of a physical quantity quite positively, whereas there is still much discussion about the extent to which these principles can be extended to other fields of study.

In the discussion which follows we restrict ourselves to scalar, as opposed to vector, quantities. The conditions required for a property of a system to be recognised as a quantity may be summarised as follows. First, the property must possess physical or theoretical significance. It is possible to invent quantities for which the following rules will apply, but which have no real meaning. For example, it is possible to assign quite accurate values to a variable which is formed by taking the product of a man's height and age (his 'hage', measured in metre.years), but such a quantity is at present of no value whatsoever. The second requirement is that it shall be possible to find procedures which allow all of the objects or events which demonstrate the property to be ordered, that is, arranged in order of magnitude.

The ordering of objects in general is quasi-serial, that is, objects are divided into groups in which each member possesses the same magnitude of the quantity, and the groups are then arranged in ascending order. To do this, we require both an equivalence and an ordering relationship, represented respectively by the symbols '$\leftrightarrow$' and '$\rightarrow$'. The statement $P_A \leftrightarrow P_B$ indicates that it may be shown that A and B possess the property P to the same degree, while $P_A \rightarrow P_B$ states that A possesses P to a greater extent than B. Although these relationships may look obvious for scalar physical quantities, there are many common properties of real objects and materials which do not obey them. If A, B and C are members of the domain $\{D\}$ of items which display the property P, for P to have the characteristics of a quantity, the following relationships must be obeyed:

Equivalence :
 (i)   reflexive $P_A \leftrightarrow P_A$
 (ii) symmetric   if $P_A \leftrightarrow P_B$ then $P_B \leftrightarrow P_A$
 (iii) transitive   if $P_A \leftrightarrow P_B$ and $P_B \leftrightarrow P_C$ then $P_A \leftrightarrow P_C$
Ordering :
 (iv) assymetric   if $P_A \rightarrow P_B$ then it is not the case $P_B \rightarrow P_A$
 (v)   transitive   if $P_A \rightarrow P_B$ and $P_B \rightarrow P_C$ then $P_A \rightarrow P_C$
 (vi) completeness   the relationships $P_A \leftrightarrow P_B$, $P_A \rightarrow P_B$ and $P_B \rightarrow$
                   P_A exhaust all the possibilities between A and B.

The next step is to establish a procedure which assigns to each member of the domain a number M which must of course obey similar relationships, that is,

$$\text{If } P_A \leftrightarrow P_B \text{ then } M_A = M_B$$
$$\text{If } P_A \rightarrow P_B \text{ then } M_A > M_B$$

For physical quantities, the density of sub-sets is extremely high, and we may map these onto part of the domain of real numbers, that is, we may treat the quantity as essentially taking a range of continuous values. The range is not necessarily the whole range of real numbers, as many quantities are limited to positive values. Although it is not common, we may select the range to suit particular applications; temperatures expressed in degrees celsius, for example, can take values above a lower limit of $-273.15\,°C$, so that everyday temperatures are small in value and usually positive.

It should be noted that these procedures do not imply any relationships between the numbers assigned in different parts of the permitted range, that is, the suggestion that such scales must be 'linear' in some way is meaningless. In addition, the magnitude of the unit interval, that is, the difference between two states whose assigned numbers differ by unity, cannot necessarily be related in different parts of the scale. This means that the quantity defined by the ordering and numbering procedures is not necessarily additive. Early theories of measurement laid great emphasis on the concepts of 'additivity' and 'unit quantities' and it is important to realise that these are not relevant to all physical quantities. Different sets of procedures may be developed for the same quantity, and give rise to a variety of scales for that quantity. It is not true, however, that similar procedures necessarily relate to the same quantity.

We have also ignored the problems raised for the ordering procedures by the presence of uncertainty in any real measurement, which means that in practice no two states are ever found to be exactly equivalent, and that we cannot even be sure that one is 'greater' than another if the measured quantities lie within the range of uncertainty. However, we may accept that the ordering procedures are to this extent idealised, and that the conclusions drawn from them are continually and more finely tested as experimental techniques improve.

A.3 EMPIRICAL SCALES

In the simplest cases, the equivalence, ordering and numbering procedures give an operational definition of a scale, that is, a recipe which describes in sufficient detail the construction and usage of measuring instruments so that values on the scale may be determined. In the discussion which follows we shall restrict the use of the term 'scale' to this type of empirical scale.

We may take as an example hardness, the property of solid materials which indicates their resistance to local deformation. An early scale was introduced by Moh in 1820 to assist in the identification of minerals. He arranged ten selected minerals, from talc to diamond, in order of hardness, assigning them arbitrary values from one to ten. The ordering relationship was given by the ability of a harder substance to scratch a softer. If neither scratches the other, they are deemed to be of equivalent hardness. This scale is clearly very primitive. The equivalence relationship is by no means exact, and it is not possible to assign fractional values, but only to say that the hardness of the sample under test lies between two of the standard minerals.

The importance of measuring hardness for a wide range of industrial materials has given rise to a variety of practical techniques and associated scales. The scratch test was extended by drawing a diamond tip over the surface of the sample under a given load, and measuring the width of the scratch produced. The Brinell hardness tester uses a 10 mm diameter steel ball which is pressed into the surface under a specified load, for example 3000 kg applied for 30 sec. The diameter of the circular indentation is measured with a microscope and converted into a hardness value. The Vickers test, on the other hand, uses a pyramidal diamond indenter. Whereas these techniques require careful preparation of the surface by polishing, the Rockwell technique measures the incremental increase in the depth of the impression when the load is increased from a light initial value to a much larger one. In fact, there are several Rockwell scales, each based on a different indenter and loading, and each suitable for a different range of materials.

Although the ordering produced in different hardness scales is usually similar, each is in fact unique, and it is not recommended that readings on one scale should be converted to, or compared with, those on another. This is due to the fact that hardness measurements involve several material properties, and the combination varies from one technique to another. In the Brinell test, for example, the pattern of stresses associated with the combination of plastic and elastic flow is complex, and changes with the depth of penetration. The elastic flow is of course recovered when the load is removed before measurement, and may distort the form of the indentation. The shape of the Brinell indenter may distort under load, also affecting the values obtained. While we might regard the differences between scales as the results of systematic errors which are difficult to assess, our understanding of the concept of hardness is limited, and we may alternatively consider that the Brinell, Vickers and Rockwell hardnesses are closely related but different quantities.

A.4 THEORETICAL QUANTITIES

While simple scales allow some advances to be made in the collection and representation of physical data, the major step forward in a scientific field is the development of an effective theory from the discovery of numerical relationships between quantities. This has several advantages. First, the relationships can be treated with the mathematical methods of representation and analysis. Since these are simply sets of rules for handling abstract quantities, they are of universal validity. Secondly, the basic relationships may suggest physical models which enable the form of the relationships to be explained and understood. Even more important is the fact that both of these processes enable new relationships to be evolved, which may then be tested experimentally. The results may confirm the validity of the theory, or may require that it should be modified or even abandoned.

The development of tried and tested scientific theories adds greatly to the significance of the fundamental physical quantities involved. It becomes possible to study them on an abstract or theoretical basis, that is, quite independently of the specific materials or procedures required for empirical scales. As the applications of a given theory extend to cover a variety of different areas, and the extent of experimental verification broadens, we become increasingly confident that the quantities common to each area have a real and general significance (although whether they reveal a universal truth or reality is a matter for philosophical debate). For example, we would consider that the same quantity 'length' is measured by comparison with a standard metre rule, by counting fringes in an interferometer, or by timing the passage of light from one extremity to the other. This does not mean, of course, that all techniques are equally good – the random and systematic uncertainties may vary from one to another – but otherwise each should give the same answer. In fact, the discovery of small differences between measurements or scales often indicates the presence of unsuspected perturbations or systematic errors. An example is provided by the period of rotation of the Earth, used at one stage to define the unit of time, whose variations could be accurately followed only after the development of atomic clocks with greatly improved accuracy.

For a quantity without a solid theoretical background, scales based upon different operational definitions will give arbitrary values which are not necessarily related in any predictable way. One of the first indications that a quantity has real physical significance and is well understood is that it becomes apparent that there are simple numerical relationships between scales based on a variety of physical effects. Often, values on one scale may be converted to another with a linear transformation. Initially, the relationships may be obscured by experimental uncertainties and the presence of unsuspected systematic errors. But eventually, when the structure of the theory is sufficiently well established, it becomes possible to jettison the empirical system of arbitrary scales, and to replace it with a system of measurement based upon a standard amount of the quantity, known as the unit quantity. Different magnitudes of the quantity may then be built up from multiples and sub-multiples of the unit quantity, using any of the established relationships within the theoretical structure. The major advantage of adopting this approach is that of generality, as the concepts involved may be applied equally to new physical laws and experimental techniques, as the subject area develops and the theoretical superstructure is extended.

The situation is particularly simple when the quantity is extensive, that is, it is proportional to the amount of material present, and the property P is additive, that is, there is a simple linear relationship:

$$P_{A+B} = P_A \oplus P_B \tag{A.1}$$

where the symbol $\oplus$ indicates the combination rule. For masses, for example, the object A would simply be added to B to combine them. The case of combining lengths is slightly more complex, as it is necessary to ensure that they are aligned in the same direction. The additivity relationship enables the construction, by means of the processes of addition and sub-division, of any multiple of the basic unit for comparison with an unknown quantity Q. We may then write

$$Q = |Q|[Q] \tag{A.2}$$

where $|Q|$ is the magnitude of the (scalar) quantity, given by a real number within the range allowed, and $[Q]$ represents the unit quantity. For completeness, it should be noted that we may also need to specify a zero or reference quantity to enable the magnitude $|Q|$ to be determined. For some quantities, for example mass, common usage requires the absolute value, whereas for others we normally deal with an interval or difference. Distances and lengths are measured from one given point to another; there is no agreed absolute reference point from which all distances are measured. In the case of time, the measurement of absolute time or date, and of time intervals or durations are both of importance. As the time origin at the beginning of the universe is totally inaccessible to us, it has been necessary to specify a reference event to act as a convenient zero from which to measure.

The adoption of a system of unit quantities removes the dependence on operational definitions of scales which must specify particular procedures and materials to be used in the measurement process, and allows the application of any of the physical laws which involve the quantities concerned. However, the point of contact with the real world then occurs in the need to be specific about the definition of the unit quantity itself. In classical physics the units or primary standards were realised in the form of material artifacts, such as the platinum-iridium metre bar and the kilogram. It was necessary to treat them with great care, and to bring them out for comparison with secondary or national standards on relatively rare occasions. Although the system was not so fragile that an accident to the primary standard would have led to a significant change in the unit quantity, it was extremely difficult to ensure that slow drifts in their absolute magnitude did not occur.

More recently some of the definitions of units have been changed so that they refer to atomic rather than macroscopic properties. Of course this has been carried out without introducing a significant change in the magnitude of the unit. The atomic processes concerned are more precisely realisable than the material standards, otherwise they would not have been chosen to replace them. They have in addition two very significant advantages. First, it is a fundamental tenet of quantum theory that atoms of a particular isotope and in a given state are not simply similar in their characteristics but identical. The unit quantity may therefore be realised independently in laboratories around the world and should in principle be the same in each. It is therefore much more accessible than a single artifact, and much less prone to accidental change. In practice, of course, the distinction is blurred by the effects of systematic uncertainties which differ from one laboratory to the next. This leads to the establishment of systems for intercomparing the national standards in order to assess the overall reproducibility of the unit. The second advantage is that there is no reason to suppose that atomic properties vary with time. Although some theories suggest that fundamental constants may change on a cosmological time scale, the effects are limited, very small in magnitude and have not been confirmed in practice.

A given quantity may of course be measured in terms of a number of different units, either for convenience in measuring quantities of widely differing natures or magnitudes, or simply as a result of cultural or traditional influences. The units are related by multiplying or conversion factors, so that a measurement in terms of one unit may readily be converted into another. The need for standardisation has resulted in the recommendation of an international system of units, known briefly as the SI units. In this, a division is made into base, derived and supplementary units. The base units are those of length, the metre; mass, the kilogram; time, the second; electrical current, the ampere; thermodynamic temperature, the kelvin; the quantity of material, the mole; and of luminous intensity, the candela. Each of these is accompanied by a definition which effectively determines the magnitude of the unit quantity by assigning a number to a specified object or process. In some cases, the means of realising the unit is also indicated. The base units are of course related to the system of fundamental constants. For example, the unit of length was until recently defined in terms of an assigned value for the wavelength of a spectral line of krypton. It has now been related directly to the unit of time by giving a fixed value to the speed c of light *in vacuo*. The metre is now defined as the distance travelled by light in a time $1/c$.

In the case of the derived units, that is, those which are formed by combinations of the base units, it is undesirable to lay down procedures for their definition, as this might lead to inconsistency. Depending upon the methods used to measure these quantities, it may however be of value to establish secondary standards in order to maintain or disseminate the quantity. One example is the Weston cadmium standard cell, which is used as a working standard for maintaining the volt. The supplementary units include those of angle and solid angle, and those associated with specialised areas, such as radioactivity.

We might ask whether our ideas based on additive quantities may be extended to those cases where simple additivity is not possible, in particular for intensive quantities (that is, those which do not depend upon the amount of material present, such as pressure and density). In these cases the combination rules are likely to be indirect and more complex. For derived quantities, we cannot, for example, add two liquids of the same density to obtain one with twice the density. But we can, by specifying the condition that equal volumes are taken, relate the densities to the weights of the liquids. For these quantities, it is necessary to employ our theoretical understanding of the concept in order to suggest valid combination rules. And because of these, the statement that "the magnitude of the quantity associated with object A is twice that of B" is also meaningful for derived and intensive quantities.

It is important to realise that the same statement is not correct for a quantity measured on an empirical scale. In the absence of a theoretical knowledge of the physical basis for the scale – and if this exists the scale is in many cases unnecessary – it is not possible to relate values at different points in a sensible fashion. Although it is possible in general to identify a point which has twice the numerical value of another, this has no significance beyond the arbitrary relationship within the scale definition which led to the assignment of those values. It is commonly assumed that all measurements should be quoted in the form of (A.2) above. On the basis of the arguments above, it might be suggested that an alternative equation should be recognised for scale quantities:

$$Q = |Q|[Q\$] \tag{A.3}$$

where now $|Q|$ represents the numerical position on the scale, and $Q\$$ is a short string of symbols which identify the scale. However, it is not unusual for quantities measured on an arbitrary scale to be used in the determination of derived quantities, and it is clear that we must allow $Q\$$ to take on some of the characteristics of a unit. We may refer to it as a 'unit scale interval'; it should be stressed that there is no exact equivalence in any non-trivial way between unit scale intervals at different parts of the scale. Because the basis of the scale is empirical, derived quantities will, if for no other reason, tend to depend on the position on the scale at which the measurements are made. To summarise, the 'unit quantity' and the 'unit scale interval' are fundamentally different concepts, however much trouble is taken in attempting to ensure that their magnitudes are similar in practice.

Although quantities measured in terms of an empirical scale do not possess unit quantities, it is not always true that those measured in terms of a proper unit do not require a scale. For example, the time scale UTC (Universal Coordinated Time) has a very precise unit, the second, based upon the microwave transition used in a caesium beam atomic clock. It was brought into agreement with atomic time at 0 hr 0 min 0 s on January 1st, 1972, effectively setting the zero of the scale. But to minimise the difference which might accumulate with a time scale based upon the rotation of the Earth, which is not uniform, discontinuities of one second – 'leap seconds' – may be (or may not be) introduced at 6-month intervals. It follows that time intervals on UTC of the same numerical magnitude are not always of the same real duration. The practical consequences are small and usually negligible, but the example does show the problems that may arise if we try to impose an over-simplistic structure onto real practice.

Effective wavelengths

B.1 MEAN EFFECTIVE WAVELENGTH

Before the availability of cheap computing power a mainframe computer was required to calculate integrals precisely, and it would have been extremely time consuming and inconvenient to carry out the process of solving Planck's equation numerically for each measurement. The solution adopted, which is still quite commonly used today, was to find a simple analytical function which could represent the signal ratio $I(T)/I(T_0)$.

The function employed is an 'effective wavelength', that is, a parameter close to the centre wavelength of the radiation thermometer's wavelength function $\Phi(\lambda)$ but which could vary to some extent with the temperature T of the source. We have noted before that the equation

$$M(\lambda, T) = c_1 \lambda^{-5} \exp(-c_2/\lambda T) \tag{B.1}$$

is a good initial approximation. If the ill-defined parameter λ in this is replaced by the mean effective wavelength λ_m between the temperatures T and T_0 the equation

$$\ln I(T) - \ln I(T_0) = \frac{c_2}{\lambda_m} \left(\frac{1}{t_0} - \frac{1}{T} \right) \tag{B.2}$$

is exact. The adoption of the Wien approximation on the right-hand side of this equation greatly simplifies both the theory and the subsequent calculation of the temperature using effective wavelength values. It should be stressed that this does not in itself introduce any error into the method; it is a convenient and sensible but otherwise arbitrary choice, and other functions could have been taken.

λ_m may of course be obtained from (6.5) and (B.2) by evaluating the integrals for known values of T and T_0 using the measured characteristics of the pyrometer. The presence of the emissivity terms, generally only known with moderate accuracy for a few materials, causes some problems, which can be solved by ignoring them and then T and T_0 are treated as radiance or brightness temperatures, as was notably done in calibrating tungsten ribbon lamps as standard sources. Thus the radiance temperature T_R is the temperature of a black-body which gives the same radiance as the source at the wavelength specified:

$$\epsilon(\lambda, T)P(\lambda, T) = P(\lambda, T_R) \tag{B.3}$$

Even so, λ_m is a function of both T and T_0, and some further simplification appears to be necessary. A limiting effective wavelength λ_e which is a function of T alone is therefore defined by allowing T_0 to approach T:

$$\frac{1}{\lambda_e} = \lim_{T_0 \to T} \left(\frac{1}{\lambda_m} \right)$$

It is not in fact calculated from λ_m, but more directly from an integral equation which we will now derive. Equation (B.2) may be rearranged to give

$$-\frac{c_2}{\lambda_m} = \frac{\ln I(T) - \ln I(T_0)}{1/T - 1/T_0}$$

Taking the limit as $T_0 \to T$,

$$\begin{aligned} -\frac{c_2}{\lambda_m} &= \frac{d \ln I(T)}{d(1/T)} \\ &= \frac{1}{I(T)} \frac{dI(T)}{d(1/T)} \end{aligned} \tag{B.4}$$

As we have removed the emissivity by using radiance temperatures, the only temperature-dependent term within the integral is the Planck function $P(\lambda, T)$, and hence

$$\frac{dI(T)}{d(1/T)} = -c_2 \int_0^\infty \frac{P(\lambda, T)t(\lambda)s)(\lambda}{\lambda[1 - \exp(-c_2/\lambda T)]} d\lambda \tag{B.5}$$

At this point the Wien approximation can be used for the differentiation within the integral, which is equivalent to neglecting the term in braces in the denominator. The final expression for calculating the limiting effective wavelength is, therefore,

$$\frac{1}{\lambda_e} = \frac{\int_0^\infty \lambda^{-1} P(\lambda, T)t(\lambda)s(\lambda)d\lambda}{\int_0^\infty P(\lambda, T)t(\lambda)s(\lambda)d\lambda} \tag{B.6}$$

Having calculated λ_e from this equation, it could be presented as a table covering the temperature range of the pyrometer, or, more conveniently, by an analytical function of T. For many narrow-band pyrometers, it was found that a very close approximation for λ_e was given by

$$\frac{1}{\lambda_e} = a_0 + \frac{a_1}{T} \tag{B.7}$$

To recover the mean effective wavelength required for the calculation of temperatures, (B.4) could be integrated to obtain

$$-c_2 \int_{1/T_0}^{1/T} \frac{1}{\lambda_e} d\left(\frac{1}{T}\right) = \ln I(T) - \ln I(T_0) \tag{B.8}$$

which from (B.2)

$$= \frac{c_2}{\lambda_m}\left(\frac{1}{T_0} - \frac{1}{T}\right)$$

When the limiting effective wavelength was accurately given by (B.7), the integral could be easily evaluated, so that

$$\frac{1}{\lambda_m} = a_0 + \frac{a_1}{2}\left(\frac{1}{T_0} - \frac{1}{T}\right)$$
$$= \frac{1}{2}\left(\frac{1}{\lambda_{e0}} - \frac{1}{\lambda_e}\right) \tag{B.9}$$

that is, the mean effective wavelength between temperatures T and T_0 is simply the mean of the limiting effective wavelengths at these temperatures. It was one of the significant advantages of the effective wavelength approach that (B.9) appeared to be a very good approximation for many narrow-band pyrometers. Deviations only occurred for pyrometers with broad bandwidths and with transmission curves which were asymmetric.

While the accuracy achieved with this method was quite sufficient for many purposes, it may be inadequate for high precision thermometry. The assumption of the Wien approximation in (B.6) for λ_e leads to an error which is by no means negligible at high temperatures. As an example, the limiting effective wavelength has been calculated for the model pyrometer. The values λ_E in Table B.1 were obtained with (B.6) and those for λ_C with the corrected integral from (B.5). It will be seen that the differences between the two are small at low temperatures but increase rapidly towards the top of the range, reaching 0.19 nm at 2400 °C. This is at the level of the overall uncertainty achievable and has to be considered. Less obvious from the table is the fact that λ_E obeys the linear relation in (6.9), with a maximum deviation from the least-squares linear fit of 0.005 nm at the extreme ends of the range shown. The corrected points λ_C lie on a line which is distinctly curved at low values of $1/T$, that is, at high temperatures. Clearly, the simplicity and efficiency of the effective wavelength method would be lost if it were necessary to use λ_C in order to achieve high accuracy.

A solution to the problem is to change the definition of the mean effective wavelength in (B.2) to use the Planck function rather than the Wien approximation $W(\lambda, T)$, that is,

$$\ln I(T) - \ln I(T_0) = \ln P(\lambda, T) - \ln P(\lambda, T_0) \tag{B.10}$$

Table B.1: Comparison of limiting effective wavelengths for our model radiation thermometer calculated by different methods.

	By integration		By series expansion	
T/°C	λ_W (B.6)	λ_P (B.5)	Approximate (6.20)	+ Extra terms (6.21)
800	662.446	662.446	662.506	662.446
1000	661.679	661.679	661.716	661.680
1200	661.119	661.119	661.142	661.120
1400	660.692	660.690	660.706	660.692
1600	660.355	660.349	660.364	660.355
1800	660.082	660.064	660.087	660.082
2000	659.857	659.812	659.860	659.857
2200	659.669	659.570	659.669	659.668
2400	659.508	659.318	659.507	659.507

then the left-hand side of (B.4) becomes

$$\frac{d \ln P(\lambda_e, T)}{d(1/T)} = \frac{1}{P(\lambda_e, T)} \frac{dP(\lambda_e, T)}{d(1/T)}$$
$$= -\frac{c_2}{\lambda_e} \frac{1}{1 - \exp(-c_2/\lambda_e T)} \quad \text{(B.11)}$$

The term within the braces in the denominator of (B.5) is close to unity and changes very little over the bandwidth of the pyrometer. It may therefore be removed from within the integral, and it then cancels with the similar term in (B.11) above. The error in this procedure is very small; for the model pyrometer it reaches a few parts per million at the highest temperatures. The overall conclusion is slightly surprising; to avoid both the error and the problems of curvature if the corrected integral in (B.5) is used, the limiting effective wavelength should continue to be calculated with the conventional expression in (B.6), but the mean effective wavelength obtained from it should be substituted into an equation using Planck's function rather than the Wien approximation, that is, the mean effective wavelength is defined by, and used in, (B.10) rather than (B.2).

One point arising from this remains. Since the Wien approximation was also employed in (B.4) from which the relationship between the mean and limiting effective wavelengths was derived, we may ask whether this relationship is still valid. The answer, obtained from some reasonably tedious calculations which will not be presented here, is that the error introduced is of the order of a few parts in a million, and is therefore quite negligible even for the most precise measurements. The simplicity of (B.9) for calculating the mean effective wavelength for many pyrometers is therefore retained. It should be noted, however, that the first line of (B.8) is no longer exact. From (B.11) we may derive

$$\ln I(T) - \ln I(T_0) = -c_2 \int_{1/T_0}^{1/T} \frac{1}{\lambda_e} \left[1 - \exp\left(-\frac{c_2}{\lambda_e T} \right) \right]^{-1} d\left(\frac{1}{T} \right)$$

$$= -c_2 \int_{1/T_0}^{1/T} \frac{1}{\lambda_e} d\left(\frac{1}{T} \right) + \left[\exp\left(-\frac{c_2}{\lambda_m T} \right) - \exp\left(-\frac{c_2}{\lambda_m T_0} \right) \right] \quad \text{(B.12)}$$

The wavelength λ_m in the exponential correction terms is not required with any accuracy, as both are small.

B.2   OTHER EMPIRICAL EQUATIONS

One weakness of the effective wavelength approach is that it fails to make explicit use of the fact that the signal $I(T)$ from a particular pyrometer is a function of T only, so that it is possible to write

$$\ln I(T) - \ln I(T_0) = F(T) - F(T_0)$$

which is symmetrical with respect to the interchange of T and T_0. It is therefore only necessary to find a single function $F(T)$ which may contain an arbitrary additive constant, to take care of the dependence of the signal ratio on both T and T_0. We may recall that this problem led to the rather confusing system of mean and limiting effective wavelengths described in the previous section.

For example, if we generalise (B.7) to include those pyrometers which do not follow the simple 'linear' relationship, we may write

$$\frac{1}{\lambda_e} = \sum_{i=0}^{i=N} a_i \frac{1}{T^i} \quad \text{(B.13)}$$

and substitution into (B.8) yields directly

$$\ln Q = \ln \frac{I(T)}{I(T_0)} = \sum_{i=0}^{i=N} a_i \left(\frac{1}{T_0^{i+1}} - \frac{1}{T^{i+1}} \right) \quad \text{(B.14)}$$

The coefficients a_i may be obtained by fitting (B.13) to the calculated limiting effective wavelength, or more directly, by fitting a polynomial in $1/T$ to $\ln Q$. In the latter case, the fitting will take into account the small exponential terms in (B.12). The errors introduced by this procedure are indicated by the magnitude of the residuals from the fitting procedure, and it is good practice to establish that the fit with the degree N of the polynomial selected is adequate throughout the operating range of the pyrometer by including temperatures at $50\,°C$ or $100\,°C$ intervals. If the residuals are unacceptably large, for broad filters for example, N should of course be increased. There is no point in deriving unnecessarily precise fits; the maximum residual should correspond to a temperature error around three to five times less than the total uncertainty required at that temperature. The solution of (B.14) may then be obtained by the same iterative procedure described in Chapter 6, with the iterative approximation shown in equation (6.7).

The value of these empirical equations is that they enable the performance of a pyrometer to be summarised with a small number of parameters. They avoid the need to retain the considerable volume of measurements needed to fully represent the filter transmission curve $t(\lambda)$ and the spectral responsivity $s(\lambda)$ in the integrals of (6.5) and (6.6).

Measurement of filter transmission

C.1 INTRODUCTION

It was demonstrated in Chapter 6 that, whichever technique is used to process results obtained by primary radiation thermometry, a precise knowledge of the detailed shape of the transmission curve $t(\lambda)$ of the filter, or of the spectral responsivity of the complete instrument $\Phi(\lambda)$, is essential. Because the results consist of signal ratios, only relative values of $t(\lambda)$ or $\Phi(\lambda)$ are required, and arbitrary scaling factors may be used as convenient. For the most accurate thermometry, the uncertainties in the measurement of $t(\lambda)$ or $\Phi(\lambda)$ are often a major factor in the overall uncertainties achieved. In this section, therefore, we shall look at the methods employed and the sources of error in them. For secondary calibration thermometers, a rough approximation to the shape of $t(\lambda)$ is often sufficient, while industrial thermometers need the operating wavelength only crudely, for the estimation of the emissivity of a surface, for example.

While fundamental metrologists would undoubtedly prefer to measure the factors $\tau(\lambda)$, $t(\lambda)$ and $s(\lambda)$ separately, it is not always possible to do so. Because the characteristics of interference filters depend to some extent on the area illuminated by the beam and its angle of incidence, precise measurements must be made with the filter in place in the instrument. To determine $t(\lambda)$ alone, it is necessary to repeatedly move the filter out of the optical path and replace it exactly in its original position. If the elements of the radiation thermometer are fixed, only the spectral responsivity of the instrument as a whole can be measured.

A basic measurement system is shown in Figure C.1. Monochromatic radiation is produced by a combination of a continuum source such as a gas-filled strip lamp or quartz iodine lamp and a tuneable high-resolution monochromator. A diffuser or an integrating sphere may be necessary after the exit slit to provide a uniform target area for the thermometer, and to remove any polarisation induced by the monochromator. To avoid the need to know the properties of the lamp and monochromator, the intensity of the monochromatic beam produced is monitored

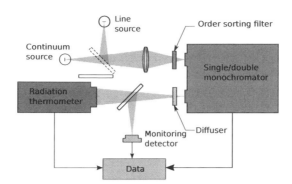

Figure C.1: Schematic of apparatus used to measure $t(\lambda)$.

either with a detector whose spectral responsivity $s_M(\lambda)$ has been measured, for example a large-area silicon photodiode in the visible or near-infrared region, or one whose responsivity is flat over the wavelength range required.

To determine $\Phi(\lambda)$, the ratio of the signal $I(\lambda)$ from the radiation thermometer detector to that $I_M(\lambda)$ from the monitoring detector is recorded as the monochromator is stepped through the wavelength range to which the thermometer is sensitive. It is given by

$$\Phi(\lambda) = s_M(\lambda)\frac{I(\lambda)}{I_M(\lambda)}$$

As the shape, rather than the absolute value, is required, no corrections are necessary for multiplying or scaling factors such as the field of view of the detector, as long as their relative values do not change with wavelength. It is also assumed here that the detectors are linear, or may be corrected for non-linearity. This is not a major problem for the monitoring detector, as the intensity of the radiation from the monochromator will not vary greatly over a limited wavelength range. However, the radiation thermometer signal will vary over several decades from the wavelength of maximum transmission out into the wings, and the effects of non-linearity must be considered.

To make separate determinations of $\tau(\lambda)$, $t(\lambda)$ and $s(\lambda)$, the range and colour filters are first removed from the optical path. The spectral responsivity of the remaining components, mainly from the detector since the transmission of the optical system should be flat, can then be obtained by direct comparison of the thermometer signal with that from the monitoring detector. It is only necessary to make measurements at widely spaced wavelengths, for example 20 nm apart in the visible and near infrared, as $s(\lambda)$ should be smooth. Similarly, the transmission curves of the range filters may be checked over the range of interest. In both cases, the variation with wavelength is more important than the absolute values of $s(\lambda)$ or $\tau(\lambda)$, although estimates of these serve to check the correct performance of the thermometer. The transmission curve of the colour filter can then be obtained from the ratio of the detector signals with the filter in and out of the optical path. Absolute values of $t(\lambda)$ are obtained with this technique, and are of use in monitoring the long-term stability of the filter.

The source of radiation needs to be bright, as the transmission of the monochromator and diffuser together are low, especially if the slit widths are small to minimise the spectral bandwidth. If the detector signals are measured simultaneously or within a short interval, no great stability for the source is required. Under some conditions, where good accuracy over moderate times is required, the source may need to be stabilised, either by controlling the current through the lamp or with feedback from a photodetector monitoring the output radiation.

For the most accurate measurements, the monochromator itself must be selected with care. The spectral bandwidth of the radiation reaching the radiation thermometer should be no more than one-tenth that of the colour filter. In some cases, monochromators may be modified to give a double-pass configuration with a reduced bandwidth, but at the cost of significantly reduced throughput. The output should of course be free of stray radiation and 'ghosts' at wavelengths other than that required. Glass filters with a range of sharp cut-off wavelengths (for example, the Schott RG series) are valuable for detecting the presence and level of undesired radiation in the output beam.

Variation of the wavelength passed by the monochromator is usually achieved by rotating the grating via a precision screw-thread arrangement driven by a stepper motor. The wavelength is indicated either by a vernier scale on a large drum which can also be used for manual adjustments, or by a mechanical or electronic counter which records the number of steps taken. It is of course an advantage if the relationship between the wavelength and the reading is linear, but in any case the reading must be calibrated with a source of narrow spectral lines. Small discharge tubes containing neon or krypton, for example, provide a large number of narrow lines whose wavelengths are accurately known. It should be noted that the values given in most reference handbooks are the 'air wavelengths', and should be multiplied by the refractive index of air (1.00028 in the visible and near infrared) to convert the scale derived into wavelengths *in vacuo*. Any eccentricity in the screw thread may be determined by measurements at several wavelengths, but only needs to be checked occasionally. Hysteresis and backlash in the mechanical couplings can be evaluated by stepping through a spectral line first in one direction and then the other. It is good practice to make a complete set of measurements including those of the calibrating spectral lines in the same direction. To simplify this procedure, the wavelengths of filters can be selected to lie centrally between convenient pairs of spectral lines.

C.2 MEASUREMENT PROCEDURES

Many readings are required in order to characterise the spectral response of a radiation thermometer completely. To minimise the effects of drifts in the detectors or in the temperature of the filter, it is desirable to allow the system to come to equilibrium after switching on within a stable environment, and to keep the total measurement time as short as possible. The measurement procedure should be carefully organised to reduce the total time required. In particular, the distribution and number of readings should be selected to match the shape of the transmission curve and the numerical integration methods to be used.

The first step therefore is to gain an overall impression of the shape of $t(\lambda)$ and the total wavelength range to be covered, by making measurements at rather wide intervals and with moderate accuracy. At this stage, filters with poor characteristics, for example with sidebands of significant amplitude, greater than 10^{-4}, or background transmission greater than 10^{-5}, may be weeded out. The range of wavelengths detected by the radiation thermometer is limited in the absence of the filter by the fall in the radiance from the Planck distribution at short wavelengths, and by the cut-off of the detector response at long wavelengths. With the filter in place, the contribution to the overall response becomes negligible well within these limits, but it is necessary to survey $t(\lambda)$ well away from the main peak to ensure an accurate representation. It is not unusual to find small sidebands in these regions, but it must be confirmed that these are not 'ghosts' of the main peak produced by defects in the monochromator. Ghosts and stray radiation may be studied by illuminating the monochromator with a purely monochromatic source, for example a laser with an interference filter which removes non-lasing lines and continuum radiation, and observing the throughput at other wavelengths.

It is often assumed that the data points should be concentrated around the peak of $t(\lambda)$. There are two reasons why this should not be the case. First, in determining the most important parameter, the mean wavelength λ_0, data points are effectively weighted in proportion to their distance from the centre. That is, an error well away from λ_0 has a greater effect than the same error at the centre. Secondly, the contribution to the uncertainty in the integration of a particular segment of $t(\lambda)$ is proportional to the curvature there. For example, if the simple trapezium rule is employed for integration, the error from an interval of width h is given by

$$\delta_i = \frac{1}{12} h^3 \frac{d^2 t(\lambda)}{d\lambda^2}$$

The density of data points should therefore be highest where the slope of the curve changes quickly, and least where the transmission is low and changing slowly, i.e., in the wings. Since the transmission in the wings may fall to a few parts per million or less, it may be necessary to increase the beam intensity in order to produce a measurable transmitted signal. As the curve is generally very flat in this region, the monochromator slit widths may be increased for this purpose without introducing significant error. Because the transmitted signal is very low, it is essential that no stray light, from inter-reflections or other sources, should reach the detector.

The determination of $t(\lambda)$ for a given filter will therefore consist of a set of measurements taken at the selected wavelengths in one direction, and including the spectral line calibration of the wavelength scale, followed by a similar set in the reverse direction. This procedure compensates to some extent for slow drifts in factors which affect the filter characteristics. Each data point may consist of several readings, averaged to give the required statistical accuracy. At this stage, each reading is checked for consistency with adjacent values to ensure that extraneous noise or interference is not present. The temperature coefficient of the filter may also be determined by repeating some of the readings at the peak and on the sides at different ambient temperatures. Small temperature changes may be accounted for in terms of the changes in the peak transmission t_0 and the mean wavelength λ_0.

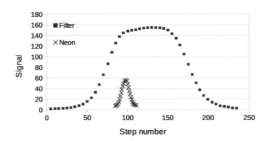

Figure C.2: Part of a data set taken at NPL during the measurement of $t(\lambda)$ for a narrow bandwidth filter, including one of the calibrating spectral lines.

Figure C.2 shows part of a data set taken at the National Physical Laboratory (NPL) for a narrow (1 nm bandwidth) 660 nm filter, using a 1-metre grating monochromator. The minimum step size was 0.01 nm and was reproducible to within 0.1% over the wavelength range tested, from 550 nm to 900 nm. The observed width of the neon 659.895 nm and 667.83 nm spectral lines used for the wavelength scale calibration was 0.11 nm. The position of the centre of the lines reproduced to within one step when the sweep direction was reversed.

C.3 INTEGRATION TECHNIQUES

Whichever method is to be used to analyse the results, the first step after the determination of $t(\lambda)$ or $\Phi(\lambda)$ involves its integration in one of (6.4), (6.8) or (6.15). The raw data consist of the transmission and responsivity values at the selected wavelengths, and a weighting factor which is inversely proportional to the estimated uncertainty. If necessary, $s(\lambda)$ and $\tau(\lambda)$ may be interpolated over the range required, as they are usually measured at only a few wavelengths, by curve-fitting an appropriate function to the data.

The simplest integration technique is the trapezium method, in which the contribution to the integral from two adjacent points is simply the area of the trapezium bounded by the points themselves and their projections on the horizontal axis. The method is robust, being insensitive to random errors in the data points, and does not require the points to be evenly spaced. It is however relatively inefficient as it is a straight line approximation to a segment of a curve, and the error from each interval of width h is proportional to h^3. An estimate of the total uncertainty may be obtained by repeating the integration with alternate points omitted, when the error should increase by a factor of 4.

Fitting a curve through the data points can give a better representation of the curve and reduce the overall uncertainty (or, alternatively, give the same uncertainty with fewer points) as the error is proportional to h^5. However, this must be done with care; fitting polynomials exactly through small sets of data points can lead to significant errors, due to the random errors in the data points, and exacerbated if spacing is not uniform.

A near-optimal technique in terms of smoothness and the representation of the data employs least-squares curve-fitting with cubic splines. To provide critical assessment of the fitting adequacy and the placing of the 'knots', that is, the junction points between adjacent regions separately fitted with cubic splines, the variation from one region to another provides a check on the correctness of the distribution of the data points over $t(\lambda)$. By omitting sets of points from the ends of the measured range, it is possible to tell whether the wavelength range covered was sufficient. If the result changes significantly, the range should be increased, while, on the other hand, if no change is observed, the range was probably too wide. Ideally, a small change of the order of the required accuracy should be seen.

The coefficients and knots of the cubic splines provide an excellent representation of $t(\lambda)$ or $\Phi(\lambda)$ and form a much smaller data set than the measurements themselves. They may therefore be used to evaluate the various integrals more quickly and efficiently than can be done with the raw data, but without the inherent (if small) errors that are associated with the use of empirical equations.

C.4 CORRECTION FOR MONOCHROMATOR BANDWIDTH

A systematic error which has not been discussed so far is produced by the finite bandwidth of the radiation leaving the exit slit of the monochromator. As a result the measured transmission curve of the filter $t_m(\lambda)$ is the convolution of the transfer or slit function of the monochromator $s(l)$ and the true transmission curve $t(\lambda)$, that is,

$$t_m(\lambda) = \int_{-L_0}^{+L_0} s(\lambda)t(\lambda - l)dl \tag{C.1}$$

In practice the limits of integration $\pm L_0$ need cover only a very limited wavelength range, as $s(\lambda)$ is narrow. It will be assumed that $s(\lambda)$ is normalised so that the total transmission is unity, that is,

$$\int_{-L_0}^{+L_0} s(\lambda)dl = 1 \tag{C.2}$$

In general the changes produced in the shape of the transmission curve do not constitute a major problem. Nevertheless, it is necessary for precise applications to be able to estimate their magnitude and, if necessary, to make an appropriate correction for them. In particular, if the intensity of the source formed by the lamp, monochromator and diffuser is low, when the monochromator is used in the double-pass configuration, for example, to reduce the fraction of stray light, it may be necessary to increase the width of the entrance and exit slits to obtain an acceptable signal-to-noise ratio from the detector.

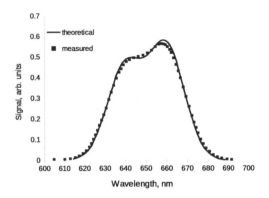

Figure C.3: Effect of broad slit function on the measured shape of a filter transmission curve.

Figure C.3 demonstrates the effect of a very broad slit function upon the observed filter transmission curve. In principle $s(\lambda)$ acts as a filter, smoothing out the more rapid changes in $t(\lambda)$. It decreases $t(\lambda)$ when the curvature is negative, that is, concave when viewed from the x-axis, and *vice versa*. While the bandwidth of the filter does not appear greatly altered, the effect of the slit function is to flatten the central peak and increase the transmission in the wings. Although the change in the wings is not very apparent in Figure C.3, when expressed as a percentage it is quite significant.

The standard approach to deconvolution, the calculation of $t_0(\lambda)$ from $t(\lambda)$, involves the application of Fourier transform theory or the use of iterative numerical methods. In radiation thermometry, we may make use of the fact that the bandwidth Δs of the slit function is always much less than that $\Delta \lambda$ of the filter or pyrometer being measured, and we may employ the Taylor series expansion technique again. Expanding $t(\lambda - 1)$ about λ,

$$t(\lambda - 1) = t(\lambda) + l\, t'(\lambda) + \frac{1}{2}l^2 t''(\lambda) + \ldots$$

where the primes indicate differentiation with respect to wavelength. Substituting this expression into the convolution integral (C.1) gives

$$t_m(\lambda) = t(\lambda) \int_{-L_0}^{+L_0} s(\lambda)dl$$
$$+ t'(\lambda) \int_{-L_0}^{+L_0} l\, s(\lambda)dl + t''(\lambda) \int_{-L_0}^{+L_0} l^2 s(\lambda)dl + \ldots$$

As l is referred to the centre of the slit function, the integrals are identical to those of (6.15) and give the moments of $s(\lambda)$ about its centre. In general the slit function is symmetrical about its centre so that the odd moments are zero and, from (C.2), the first integral is unity. Hence

$$t_m(\lambda) = t(\lambda) + \frac{1}{2} A_2 \Delta s^2 t''(\lambda) + \ldots \qquad \text{(C.3)}$$

where A_2 has the same meaning as previously, and may be taken for simple shapes from Table 6.1. For a monochromator with equal entrance and exit slit widths, we would expect $s(\lambda)$ to be a symmetrical (isosceles) triangle. It may most readily be obtained from the shapes of the spectral lines used for the wavelength scale calibration of the monochromator.

For filters with simple shapes the second differential $t''(\lambda)$ may be calculated by differentiation, but for real filters it must be derived from the measured data. This is not as simple a process as it might appear. First, numerical differentiation accentuates the effects of random noise; it is important therefore not to use the measured data, but the smoothed representation of $t(\lambda)$. If polynomials or cubic splines were used for this purpose, differentiation is a straightforward process. Secondly, the filtering effect of the slit function described above removes the higher frequency components of $t(\lambda)$ and the estimate of $t''(\lambda)$ from the measured curve may introduce a significant error. Equation (C.3) should therefore primarily be employed to determine whether the distortion introduced by the slit function leads to a significant systematic error. Its use to correct that error should only be performed with caution and preferably checked with other measurements. If data values are available or can be generated from the curve-fitting functions at uniform intervals h, then the second differential at a given wavelength can be calculated from adjacent values from the simple expression

$$t'' = \frac{t_{-2} - t_{-1} - t_{+1} + t_{+2}}{3h^2}$$

Bibliography

[1] M. P. Chappuis. Etudes sur le thermomètre a gaz et comparison des thermomètres a mercure avec le thermomètre a gaz. *Travaux et memoires du Bureau International des Poids et Mesures*, Tome VI, 1888.

[2] H. Preston-Thomas. The International Temperature Scale of 1990 (ITS-90). *Metrologia*, pages 3–10, 1990.

[3] Comité Consultatif de Thermométrie. Procès-verbaux du Comité International des Poids et Mesures. Technical Report 78th meeting, 1989.

[4] Comité Consultatif de Thermométrie. `http://www.bipm.org/en/committees/cc/cct/guide-its90.html`, 2015.

[5] Comité Consultatif de Thermométrie. *Supplementary Information for the ITS-90*. BIPM, 1997.

[6] Comité Consultatif de Thermométrie. *Techniques for Approximating the International Temperature Scale OF 1990*. BIPM, 1997.

[7] D. C. Ripple, R. Davis, B. Fellmuth, J. Fischer, G. Machin, T. Quinn, P. Steur, O. Tamura, and D. R. White. The roles of the mise en pratique for the definition of the kelvin. *International Journal of Thermophysics*, 31(8):1795–1808, 2010.

[8] H. P. Baltes and F. K. Kneubuhl. Thermal radiation in finite cavities. *Helv. Phys. Acta*, 45:481–529, 1972.

[9] H. P. Baltes. *Appl. Phys.*, 1:39–43, 1973.

[10] H. P. Baltes. *Am. J. Phys.*, 42:505–507, 1973.

[11] W. R. Blevin. *Metrologia*, 8:146–147, 1972.

[12] CODATA Task Group on Fundamental Constants. `http://www.codata.org/committees-and-groups/fundamental-physical-constants`, 2016.

[13] H. P. Baltes. *Infrared Physics*, 16:1–8, 1976.

[14] Y. S. Touloukian and D. P. ed. DeWitt. *Thermophysical properties of matter, Volume 7: Thermal radiative properties: Metallic elements and alloys*. New York: IFI/Plenum, 1970.

[15] O. S. Heavens. *Optical properties of thin solid films.* Butterworths Scientific Publications, London, 1955.

[16] E. D. Palik ed. *Handbook of optical constants of solids.* Academic Press, New York, 1985.

[17] C-D. Wen and I. Mudawar. *Int. J. Heat and Mass Transfer,* 49:4279–4289, 2006.

[18] Howard W Yoon, David W Allen, and Robert D Saunders. Methods to reduce the size-of-source effect in radiometers. *Metrologia,* 42(2):89, 2005.

[19] P. B. Coates. *High Temperatures – High Pressures,* 17:507–518, 1985.

[20] R. E. Bedford and C. K. Ma. *High Temperatures – High Pressures,* 15:119–130, 1983.

[21] M. Battuello and T. Ricolfi. *High Temperatures – High Pressures,* 12:247–252, 1980.

[22] I. Ridley and T. G. R. Beynon. *Measurement,* 7:171–176, 1989.

[23] A. Ono. Methods for reducing emissivity effects. In D. P. DeWitt and G. D. Nutter, editors, *Theory and practice of radiation pyrometry.* New York: John Wiley and Sons, 1988.

[24] Perry K. P. Drury, M. D. and T. Land. *J. Iron Steel Inst.,* 169:245–250, 1951.

[25] T. Iuchi and R. Kusaka. Two methods for simultaneous measurement of temperature and emittance using multiple reflection and specular reflection, and their application to industrial processes. In James F. Schooley, editor, *Temperature, its measurement and control in science and industry,* volume 5, pages 491–503. American Institute of Physics, New York, 1982.

[26] J. L. Gardner and T. P. Jones. *J. Phys. E,* 13:306–310, 1980.

[27] H. Kunz and D. P. DeWitt. Theory and technique for surface temperature determinations by measuring the radiance temperatures and the absorptance ratio for two wavelengths. In Anderson A. C. Janssen J. E. Cutkosky R. D. Rubin, L. G., editor, *Temperature, its Measurement and Control in Science and Industry,* volume 4, pages 599–610. Instrument Society of America, Pittsburgh, 1972.

[28] A. Levick and G. Edwards. *Analytical Sciences,* 17:s438 s441, 2002.

[29] T. Iuchi. Radiation thermometry of low emissivity metals near room temperature. In James F. Schooley, editor, *Temperature, its Measurement and Control in Science and Industry,* volume 6, page 865. American Institute of Physics, New York, 1992.

Index